WILD LIARD WATERS

Canoeing Canada's Historic Liard River

by FERDI WENGER

Wild Liard Waters
© 1998 Ferdi Wenger

Caitlin Press
Box 2387, Stn. B
Prince George, BC
V2N 2S6

All rights reserved. No part of this book may be reproduced in any form by any means without the written permission of the publisher, except by a reviewer who may quote a passage in a review.

Caitlin Press acknowledges the support of the Canada Council for the Arts for our publishing program. Similarly, we acknowledge the support of the Arts Council of British Columbia.

Page design and layout by Vancouver Desktop
Cover design by Warren Clark Graphics
Index by Katherine Plett
Photos by Ferdi Wenger
Printed and bound in Canada by Marc Veilleux Imprimeur Inc.

CANADIAN CATALOGUING IN PUBLICATION DATA

Wenger, Ferdi, 1934—

Includes index.
ISBN 0-920576-72-9

    I. Title.
PS8595.E598W54   1998C813'.54   C98-910209-2
PR9199.3.W399W54   1998

Contents

Preface / VII

CHAPTER 1
A Brief History / 1

CHAPTER 2
The Liard's Course / 3

CHAPTER 3
Falling in Love with a River / 7

CHAPTER 4
Cranberry Portage / 24

CHAPTER 5
The Voyages of John M. McLeod and Robert Campbell / 33

CHAPTER 6
French Canadian Voyageurs / 47

CHAPTER 7
In the Footsteps of the Fur Traders / 56

CHAPTER 8
The Gold Seekers / 71

CHAPTER 9
Across the Devil's Portage / 86

CHAPTER 10
Bear in Camp / 98

CHAPTER 11
Moosehunt and Rainstorms / 104

CHAPTER 12
Nahanni Butte / 122

CHAPTER 13
The End of the Trip / 131

CHAPTER 14
Grand Canyon Days / 135

CHAPTER 15
Doomed to be Dammed? / 159

Glossary / 163

Index / 164

Preface

Harvey Fraser and I stood on top of a high rocky overhang and looked down into a section of the Grand Canyon of the Liard River we had come to call "The Gates." Far below the churning waters—unusually high for early September—ran wild and angry. Awed by the wild beauty of the canyon and fascinated by the swirling, clashing waters, we sat on some rocks to study the Liard's currents.

The rocky ridge we were on was thinly covered with the tall, spearlike spruce trees of the north country. It jutted far into the Liard's boiling currents, dropping more than a hundred metres straight down. On the opposite shore the equally high grey shale of the cliffs hemmed in and forced the waters into a tight turn. The rapid constriction of the streambed, combined with the sharp bends, forced the Liard into an angry uproar. Whirlpools like giant pinwheels, easily visible even from this height, and huge boils and heavy rolling waves that rushed out of the tight gap below to the left of us, piled up against the steep walls where the river turned again. From the rocky wall foaming breakers lashed back into the stream and added their power to the general turmoil of crisscrossing currents.

As we studied the currents, trying to determine a safe passage for our canoes, a large spruce tree, dwarfed only by the height from where we saw it, came floating around the outside of the last river bend. Slowly at first, the tree spun around. It picked up speed, spun faster and faster, now caught in the edge of one of the whirlpools. Incredibly, the tree slowly stood upright, as if rigged by guide cables, and spun around several times. Then, ever so slowly, it began to disappear—still upright—sucked out of sight by the vortex.

We stared into the waves, amazed. This was our third canoe voyage on the Liard; yet the almost malevolent power of its unharnessed waters remained as fascinating to observe, and encounter, as our first. Spellbound, we watched for the tree to pop up farther downstream. But, even though we searched the waters for more than half an hour, and were able to observe more than half a kilometre of the river, the huge spruce tree remained lost in the deep, swirling jaws of the whirlpool. A vague uneasiness settled on my mind.

The words of R.G. McConnell, who first officially explored this tumultuous river back in 1887, echoed a warning in my mind as we climbed back down to the canoes: "The evil reputation of this river has not been exaggerated . . . In high water it is simply one long cascade . . . ." One long cascade!

The voyageurs, working for the Hudson's Bay Company, who had already fought their way up and down this river fifty years before McConnell's chilling forewarning, had also shown great respect for this wild river. James Anderson, a Hudson's Bay Company clerk, wrote in 1853: "They (the voyageurs) invariably on rehiring endeavour to be exempted from the West Branch (Liard)."

We were in for a challenging journey.

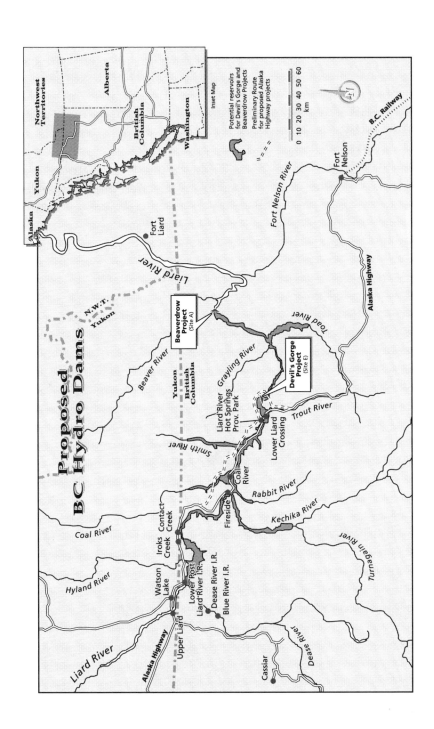

Preface

Harvey Fraser and I stood on top of a high rocky overhang and looked down into a section of the Grand Canyon of the Liard River we had come to call "The Gates." Far below the churning waters—unusually high for early September—ran wild and angry. Awed by the wild beauty of the canyon and fascinated by the swirling, clashing waters, we sat on some rocks to study the Liard's currents.

The rocky ridge we were on was thinly covered with the tall, spearlike spruce trees of the north country. It jutted far into the Liard's boiling currents, dropping more than a hundred metres straight down. On the opposite shore the equally high grey shale of the cliffs hemmed in and forced the waters into a tight turn. The rapid constriction of the streambed, combined with the sharp bends, forced the Liard into an angry uproar. Whirlpools like giant pinwheels, easily visible even from this height, and huge boils and heavy rolling waves that rushed out of the tight gap below to the left of us, piled up against the steep walls where the river turned again. From the rocky wall foaming breakers lashed back into the stream and added their power to the general turmoil of crisscrossing currents.

As we studied the currents, trying to determine a safe passage for our canoes, a large spruce tree, dwarfed only by the height from where we saw it, came floating around the outside of the last river bend. Slowly at first, the tree spun around. It picked up speed, spun faster and faster, now caught in the edge of one of the whirlpools. Incredibly, the tree slowly stood upright, as if rigged by guide cables, and spun around several times. Then, ever so slowly, it began to disappear—still upright—sucked out of sight by the vortex.

We stared into the waves, amazed. This was our third canoe voyage on the Liard; yet the almost malevolent power of its unharnessed waters remained as fascinating to observe, and encounter, as our first. Spellbound, we watched for the tree to pop up farther downstream. But, even though we searched the waters for more than half an hour, and were able to observe more than half a kilometre of the river, the huge spruce tree remained lost in the deep, swirling jaws of the whirlpool. A vague uneasiness settled on my mind.

The words of R.G. McConnell, who first officially explored this tumultuous river back in 1887, echoed a warning in my mind as we climbed back down to the canoes: "The evil reputation of this river has not been exaggerated . . . In high water it is simply one long cascade . . . ." One long cascade!

The voyageurs, working for the Hudson's Bay Company, who had already fought their way up and down this river fifty years before McConnell's chilling forewarning, had also shown great respect for this wild river. James Anderson, a Hudson's Bay Company clerk, wrote in 1853: "They (the voyageurs) invariably on rehiring endeavour to be exempted from the West Branch (Liard)."

We were in for a challenging journey.

CHAPTER 1
A Brief History

The evil reputation of this river has not been exaggerated, and it requires careful steering and hard work to navigate it with safety. It is constantly interrupted with rapids and hemmed in by narrow canyons which render frequent portages necessary. The most dangerous part of the river is included between the Little Canyon and Hell Gate. In this distance of about one hundred miles it has a fall of over a thousand feet, and in high water is simply one long cascade....
—R.G. McConnell, *Geological and Natural History Survey of Canada, 1888*

THE "GATES" IN THE GRAND CANYON of the Liard are only a small part—though one of the most beautiful and impressive—of this 1,200 kilometre long river, that rises in the Pelly Mountains of the Yukon. It is a large, beautiful and little known river. Its branches are spread out over four degrees of latitude from 58 degrees north to 62 degrees north, with one reaching within 200 kilometres of the Pacific. They interlock with the tributaries of other large rivers: the Yukon, the Skeena, the Stikine, and the Peace. From its sources in that wild and broken country of the Yukon, the Liard loops south and east through British Columbia. Upon entering the Northwest Territories it turns sharply north, to spill its waters into the Mackenzie—next to the Mississippi, the longest river in North

America—at Fort Simpson. But, as if it were reluctant to give up its personality, its identity, its muddy waters are still clearly visible for dozens of kilometres below the junction with the mighty Mackenzie on their way to the northern sea.

The Liard was once used as an avenue of trade by the men of the Hudson's Bay Company, from about 1829 when Fort Halkett was established at the mouth of Smith River, to the abandonment of that post in the 1860s. The river was first officially explored and surveyed by R.G. McConnell of the Geological Survey of Canada in 1887. He descended the Liard in a spectacular canoe journey during the season of high water, from the mouth of the Dease River. McConnell's colleague, Dr. G.M. Dawson, went upriver from the Dease and explored the river to its source that same summer.

The Liard's tortuous route was also travelled by a number of gold hungry adventurers—later known as the Klondikers—on their way to the gold fields in the Yukon in 1897 and 1898, most of them having started their journey in Edmonton. Of these brave, if sometimes foolish men, Dr. Charles Camsell was to become the most prominent when he was appointed Deputy Minister of Mines in 1920. He later wrote a book on his experiences, *Son of the North*.

Alfred E. Lee, who came up the Liard as one of the first of the gold rush, and whose experiences interacted with those of the Camsell brothers (Charles and Fred), also wrote a journal. His diary, annotated and published by Thomas L. Brock in 1976—only 165 copies have been printed—makes fascinating reading and its quotes are published here for the first time.

These men—fur traders, voyageurs, explorers, gold seekers and miners—who are all part of the history of the Liard River, often had wildly different backgrounds. After travelling the canyons of the Liard though, they all had one thing in common—a healthy respect for the perils of this river.

We knew why firsthand.

CHAPTER 2

The Liard's Course

From the mouth of the Frances River, for the first forty kilometres of its course, the Liard curves and ambles through relatively flat country, until it is swept under the Alaska Highway Bridge at Watson Lake. From there, following some lazy turns and passing a number of islands, it gathers speed. It races and leaps through the Liard Canyon, to cross the 60th parallel and to enter British Columbia. It bends to the south now, runs another twenty-five kilometres, then picks up the Dease River at a place called Lower Post.

Over the next 150 kilometres the Liard loops and twists and turns. It gathers in the Highland River, gallops through Little Canyon, where a number of Chinese gold seekers drowned. It tears and boils through more tight spots, always gathering speed to reach Cranberry Rapids. Still running parallel to the Alaska Highway and never more than a mountain ridge away, it finds no rest. Three-and-a-half kilometres below the last spray-covered boulders of Cranberry, it scoops up the Kechika River at a place that was once Mud River Post. Just downstream it tumbles over Mountain Portage Rapids, gathers in the Rabbit River and seethes through Whirlpool Canyon. Now the Liard has reached top speed. It loops north, fed by the Coal River, turns south again, only to roar and thunder past rocky walls once more, and over ledges and broken islands and past the hotpools at Portage Brûlé.

For a while the Liard quiets down, resting for yet another rumble. It flows straight and fast, past the mouth of Smith River, the site of old Fort Halkett, and then, 270 kilometres from Watson Lake, it rushes under the suspension bridge at Lower Liard. Three kilometres from the bridge is Liard Hotsprings Provincial Park.

Again the Liard gathers its forces, for the wildest ride yet. It surges past the mouth of Trout River, in a big sweeping bend, gobbling it up, and flows on to the head of the Devil's Portage. Around the narrow gorge it tightly turns back on itself foaming and boiling, and hemmed in to a mere thirty metres or less in places. Here it surges and clashes and gushes even in low water. At flood stage it deposits scores of trees on high, rocky ledges and outcrops. In low water the bleached and polished tree trunks shine white from the dark rocks and you wonder how on earth they ever got up there.

Here no boat can survive. The six-kilometre-long portage over a four-hundred-metre-high mountain ridge is the best alternative. But the Liard has just begun its rampage, for only now does it enter the Grand Canyon of the Liard, as this stretch of its concourse is somewhat fancifully named on the maps. For a distance of close to sixty kilometres it tears along, sometimes a great speed, fighting all challengers, no holds barred. Canyon follows canyon. Gate follows gate. Whirlpool follows whirlpool. Past Surrender Island. Into the Rapids of the Drowned. Through Boiler Canyon. Finally, it emerges at Hell's Gate. Its forces are spent; its battles fought.

Below the mouth of Toad River, once a lonely outpost of the fur traders, the northernmost, round, nameless peaks of the Rocky Mountains fade back. The Liard enters the lonely land of northern bush. Broad now, and braided, and crooked, its concourse is strewn with sand bars and shoals and islands. It picks up the Beaver River and then the Fort Nelson River. Turning sharply north, it spreads itself out even more and heads for Fort Liard, just a hundred kilometres away. Here at the old Hudson's Bay post it picks up the Petitot River and, keeping a general direction of north, but often almost doubling on itself in giant loops and oxbows, it grows twice its volume as the famous South Nahanni dumps its waters. Nahanni Butte, a prominent landmark, guards the junction and then bids farewell to a river that runs yet another 190 kilometres to a rendezvous with the Mackenzie. Most of that flow is smooth now. Only fifty kilometres from its mouth are there some strong riffles stretching over twenty kilometres. Only strong winds in the wide valley at the end of the river's course sometimes cause troubles for the paddler.

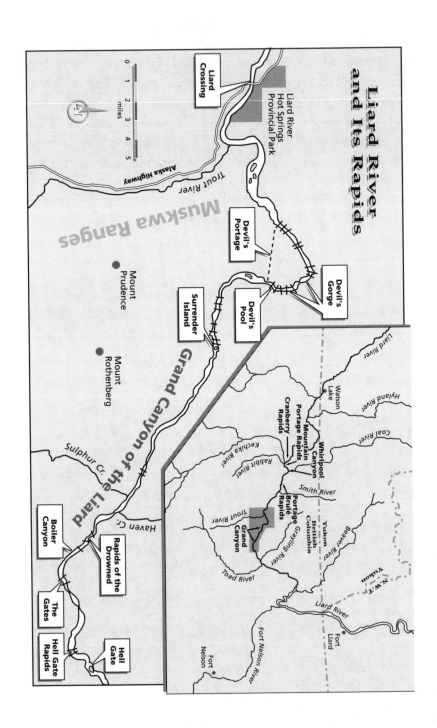

CHAPTER 3
Falling in Love with A River

> *Little Canyon is about half a mile long, and at its narrowest place about two hundred feet wide. It is easily navigable in low water, but is dangerous for small boats during flood, as the channel is very crooked, and the current striking with great violence against the right hand bank is thrown forcibly back, with the production of a number of breakers running nearly lengthwise with the direction of the channel, and large enough to swamp any ordinary riverboat which is drawn among them. A number of Chinamen were drowned at this spot some years ago....*
> —R.G. McConnell, *Geological and Natural History Survey of Canada (1888)*

I MET THE LIARD RIVER for the first time in 1970, four years before that sunny September day when Harvey and I watched the tall spruce tree disappear into the vortex of the Gates. On our first trip down the Liard we knew nothing about rafting canoes together by tying spruce poles across the thwarts to run heavy whitewater, and, to be truthful, not all that much about running whitewater. At that time, in the late 1960s, many of us were involved in Marathon Canoe Racing which was different from paddling whitewater. In marathon racing—races that could last from eight to fifteen hours—the object was speed and survival, to get from the start to finish as fast as possible, through boiling, surging currents and obstructions, and on flat water. There was no time to enjoy the rapids, to play in them, to learn the various skills. Instead it was a matter of staying upright. We relied on speed mostly, and a lot of luck, often not even wearing life jackets.

But this trip down the Liard was to be different. We had come to enjoy the river, to take our time, to enjoy the land and to explore. And in the end we fell in love with the river.

We eight paddlers who manned the four canoes all shared a love for the outdoors. Over the years we had met through work, canoeing and racing.

There was Adolf Teufele. He grew up on a farm in Bavaria but had lived in Canada for many years and was absolutely crazy about canoeing. Tall, broadshouldered and strong as an ox, also somewhat stubborn and hard-headed, he was easily the best packer in the group. A fine paddler, not easily panicked, he also was the official navigator on this the first trip.

Then there was Harvey Fraser. Unquestionably the best paddler, he was also the most experienced one. Not very tall but strong and tough, he was built like one of the early voyageurs. The colourful scarf he liked to wear as a kerchief on his head, added to the image. He also loved canoeing as they must have. A few years before, in 1967, he had—as had Adolf—taken part in that classic canoe race, the Centennial Canoe Pageant, across Canada as a member of the BC team. At that time, teams of ten men from each province and territory had followed the main routes of the voyageurs and fur traders. In a little over a hundred days, they had paddled almost 5,000 kilometres, using an eight-metre-long fibreglass canoe built along the lines of a birchbark North Canoe that the voyageurs of the fur trade had used in their trade north and west of Fort William. Harvey for many years had also ranked among the top ten marathon racers in the Northwest, races where Adolf had sometimes been his partner. Harvey grew up in the canoe paradise of northern Ontario, in the Muskoka region, and had paddled from a very early age. But best of all, he had an almost uncanny ability to read water, to pick the safest current. It is an ability that can be acquired only through experience, and so it was Harvey who led us through the wild canyons of the Liard.

Perhaps just as experienced in running whitewater was Leland Bradford. He hailed from the Kootenays, the land of mountains and rivers. Leland—called "Wink" by all his friends—was a guide, logger, rancher and white water fool. For many years Wink had guided hunting and surveying parties on horseback. That experience in the bush made him by far the best man in camp and to have around when things got tough. He could throw up a tent in less time than it took the rest of us to cut the first two poles, and cooked breakfast while the rest of us still struggled with the legs of the jeans that were forever wet.

Dale Hambleton, cool and confident, paddled with Adolf. He was American and lived in the state of Washington. A former soldier who had come through the Korean War, Dale had acquired his love for the outdoors on the rivers in a large part of the American Northwest. His great love was fishing and collecting boats. He had more canoes and kayaks at home than many of us knew existed. His passion was fly fishing and his fly rod was never far from his capable hands. The graylings that added to our breakfast always came from Dale's line.

Harvey paddled with his brother Doug. Doug, too, was an experienced canoeist, perhaps a little wiser and more cautious than the rest of the crew, he liked life jackets with lots of floatation. Together the two Frasers formed a formidable team. Whenever they fell into their long and powerful cruising stroke and steady rhythm, we all had a tough time keeping up. Worse yet, they could keep up that pace for hours. Doug had also spent a lot of time in the army. From there he had brought back his famous bannock recipe. Sitting before the simmering coals of a fire, watching the bannock slowly bake to a golden crisp, his face tanned like dark mahogany, the old hat on the back of his head, Doug always looked content. He belonged to this kind of life.

His closest rival at the campfire cooking pot was Jeff Harbottle, who came from Vancouver. At fifty years old Jeff was the oldest in the group but in excellent physical condition. He had seen much of the world in his time. He also owned an old tug boat with which he cruised the West Coast, and had run logging camps and flown his own airplane. He was a good spirit and even managed to grin when, later in the trip, he stumbled into a hornet's nest, a mistake that rendered his face an almost unrecognizable swollen mask of pain. Jeff paddled with Wink.

My partner in the canoe was Frank Hoeltschi, who made his home in Switzerland. Frank was on an extended vacation and very much impressed and surprised by the vastness of the country called Canada, by the beauty of the north and its remoteness. Things that we all too often took for granted, he still marvelled at. Thus he increased our own awareness of the beauty of nature and wildlife, that he up till now, had known only from visits to a zoo. A good paddler, Frank quickly adjusted to the new conditions of heavy water and large rivers.

Like Frank I too, had been born in Switzerland and had come to Canada in the early spring of 1956. Working for Harvey and sometimes Adolf as a carpenter's helper I had quickly been introduced to the sport of canoeing and, at the beginning, the world of marathon racing. This was my first extended canoe trip into Canada's north, one that only confirmed my love for this immense land.

This then was the crew on the first Liard trip and together we tackled the mighty river.

We reached the Liard River after a thirty-six hour drive from Kamloops in Central BC up the Alaska Highway. Wink had come up to the Yukon a few days earlier to visit his brother—a geologist and bush pilot working out of Dease Lake in northern BC. He flagged us down in Lower Post and then led us to his camp at the edge of a small lake. It was a welcome spot. We had been on the road too long, taking turns driving the dusty road, not stopping to sleep, feeling grubbier by the hour. A camp on a peaceful lake, away from the gravel road and its heavy truck traffic, looked inviting.

Later in the evening we hiked through a light stand of pines over sandy ground and across a low ridge, for a firsthand look at Liard Canyon which was to be the first obstacle that we planned to run the following day in empty canoes.

The noise of the wild, foaming rapids grew to a roar as we slid and stumbled down the last steep slope to the water's edge, grabbing for trees in a cloud of dust and sand. At water level, on the tip of a rocky tongue reaching far into the water, the noise was overwhelming. But looking upstream we knew that this rapid was more bark than bite. Thanks to low water, there would be no problems. It was—as Harvey put it—a good place to warm up for the trip.

On the opposite shore, on top of a small rocky outcrop at the end of the canyon where the turbulent waters formed a churning whirlpool, stood a simple stone monument in the memory of the great geographer and surveyor—Dr. George Dawson. Straddling the 60th parallel, it marks the border between British Columbia and the Yukon.

George Dawson had come to the Liard from Wrangell, Alaska in 1887. With him were R.G. McConnell and a group of voyageurs whose names have never been mentioned in the journals. Together the men explored the Dease River and then parted where it spilled its waters into the Liard. McConnell had continued downriver, while Dawson had gone against the current to the head of Frances Lake in the headwaters of the Liard. At the entrance to the first canyon after leaving the Dease River, where the 60th parallel crossed the river, Dawson had erected a wooden post in a pile of rocks. That post was replaced by the monument in honour of the explorer in 1936.

Liard Canyon, or Upper Liard Canyon as it was marked on some maps, was easy to run when we passed through it the first time. In high water though, it was nothing to trifle with and had claimed many victims among the careless, inexperienced travellers. This was especially true for the Klondikers who ascended it during the high water in 1897. Charles Camsell, for example, who later became the founder of the Canadian Geographic Society—one of the first to choose this difficult route to the gold fields—mentions one such tragedy in his book *Son of the North* (Ryerson Press, 1954). After having spent a few months in the vicinity of Frances Lake and having nearly starved to death, Charles Camsell and his brother Fred decided to return to Lower Post. Just below the Liard Canyon they met a group of prospectors going upstream, who fed them such huge meals that it made the two brothers ill for the next few days. Camsell wrote later in his book:

> *It was fortunate in a way for me that I did get ill from overeating, because a day or two after our arrival at Lower Post I was invited to guide a party of two men in a boat through the canyon on the Boundary Line. The men were not very experienced boat men and the canyon had some pretty rough water in it. I could not accept their proposition because I was too weak, so they went alone, one of them on shore with the tow line, and the other in the boat to steer it. Later in the evening we saw an upturned boat drifting by downstream, and, as the two men were never seen or heard from again, we came to the conclusion that they were drowned in trying to get through the canyon. Nobody at Lower Post seemed to think it was necessary to investigate this case. We were all transients. There was no police or other government authority nearer then Telegraph Creek, 250 miles away, to whom the case might have been reported, so it was simply forgotten. If I ever knew the names of these men, I have long since forgotten them. There must have been scores of cases of disappearances like this all along the route travelled by the Klondikers.*

There were no such perils for us the next day. Halfway through the morning the canoes were dancing on the whitewater of the canyon. Already the vast experience of the most seasoned paddler,

Harvey Fraser, was clearly illustrated as he guided our four canoes through the rough stuff. Almost instinctively, an instinct acquired in hundreds of kilometres of river travel, he easily picked the safest route. The cedar-strip canoes we had built the previous winter performed well. Twenty kilometres further downstream, at Lower Post, our shoulder and neck muscles not yet used to the steady rhythm of long hours of paddling, screamed for relief. Only a quick swim in the cold waters of the lake assuaged the muscular tension.

Next morning's sunrise caught us—gear and supplies now stashed in the four canoes—on the wharf of the local RCMP detachment at Lower Post. The two Indians who stood by the riverbank watching, shook their heads in bewilderment at these crazy white guys, who still paddled canoes when outboard motors and river boats were so much handier and convenient.

The Liard below the Dease flowed gently, the current fast but even. Slowly, our sore muscles warmed up and then grew accustomed to the steady rhythm. Paddling about thirty-five strokes a minute, compared to seventy or even eighty strokes a minute when racing, we could soon hold the pace hour after hour. We stopped for lunch at a sandy beach on an island where a huge pile of driftwood—five or more metres high—could have provided wood for a thousand cooking fires.

Later, we found them everywhere on the Liard, these gigantic piles of driftwood. Sometimes they were eight or ten metres high and half a kilometre long. They are witnesses to the raging flood waters of this tumultuous river. During the snow melt, in June or early July, the Liard carries thousands of uprooted trees in its dark and muddy waters. Then, when the high waters recede, most of the logs are left stranded like gigantic toothpicks—bleached white by the sun and worn smooth from their aqueous journey—on sandbars, large boulders and high rocky ledges throughout the canyons.

"For three solid days," the Mountie at Fort Liard told us later, "the Liard, here eight hundred metres wide, is a solid mass of timber. Wedged and jammed into one another, they are so thick that a man could walk from shore to shore without getting his feet wet, running along on a current of fifteen or twenty kilometres per hour."

But on that first mild and lovely day, as we cooked a luncheon soup, we were far removed from high water. We sat basking in the sun's warm rays, half asleep. Only Dale took out his fishing rod to demonstrate his skills. He flicked the fly into a small eddy behind a half-submerged log and almost instantly hooked his first grayling. Ten minutes later he had half a dozen laying on a log. Fried in butter to a golden crisp, they made a welcome addition to next morning's breakfast.

By mid-afternoon the dark clouds of a storm crested the horizon then hovered overhead. Lightning flashed and heavy, rolling thunder added gloom to the now lead-coloured river. Under pelting rain we were soon soaked to the skin. Despite the narrow strip of blue sky on the western horizon that promised perhaps evening sunshine, the heavy rain soon turned into a steady drizzle that was to last all night.

Although it was too late we dug out the rain jackets and, tucking chins in against the pummelling rain, we paddled steadily on. Forgotten for the moment, were the lazy drifting sessions we had enjoyed every hour watching the scenery. Forgotten was the beauty of the gently sloping shoreline, overgrown with willows, alders and young birches. Forgotten were the dark, wooded hills beyond, that rolled far into the distance. Instead we grimly paddled on, soon shivering violently in our wet clothes. After an hour we conceded to the storm. It was time to pull ashore, build a fire and dry out.

Just downstream, we knew from our maps and from McConnell's journal, was Little Canyon, and with it the first stretch of real rapids. Rounding the last bend, it lay before us.

Teeth chattering we turned into the current and ran the canoes onto the dark, wet sand of an island. On a small knoll stood half a dozen tall spruce trees. Their branches spread wide and protecting, they kept the ground beneath dry and offered a perfect shelter for our tent. The thick carpet of needles and moss promised a good bed. A couple of axe blazes, old and scarred, and visible only to the experienced eye, told the tale of other men, in another tent, in another time, seeking shelter and warmth at the same spot. Perhaps the men who had stopped here before us had been gold seekers, struggling upriver in the middle of winter.

The men who fought the Liard on their way to the gold fields of the Yukon had needed rest. J.G. MacGregor, in his book *The Klondike Rush Through Edmonton, 1897-1898 (McClelland and Stewart Limited, 1970)*, wrote of the first twenty-five men who set out up the Liard in the fall of 1897:

> *Some of them had been on route from Edmonton since August—over ten months. From November 7, 1897, until June the following year, they had been wandering in the wilderness from the mouth of the Beaver River a few miles below Nelson Forks, trying to complete that nine hundred mile portion of their trip to Dawson. At times they had been forced to turn back to wait better weather or better ice conditions. At times they had spent weeks on end relaying their food forward by three or four mile stages, and at times they had to divert from their direct route to purchase more supplies. If ever men struggled indomitably it was those who had ascended the Liard and who reached Dawson—those who led the way over the route which many were to use after them. And yet they were only ten out of twenty-five who had set out from Fort Simpson in 1897. Ten got to Dawson, four remained in the Cassiar for a year at least, eight went back to Fort Simpson and three died.*

Perhaps it had also been the Klondikers who had left behind the graves we had passed three or four kilometres upstream from the island. We had stopped despite the driving rain and had looked

Scouting Little Canyon with its high turbulent waters, by foot before nearing it in the canoe.

at them. The carved, white painted fence around the graves set amidst tall trees on the high bank of the river had almost fallen down, the crosses tilted and untended.

Sitting by the warm fire later in the evening, after a large and immensely satisfying supper of spaghetti and canned ham, feeling warm and comfortable despite the rain, I pulled out McConnell's report and read what he had to say about the Little Canyon, whose roar we could hear just downstream from the island.

> *This canyon can be run with safety by entering it nearly in the middle of the stream, which is as close to the left bank as the lines of reefs and isolated rocks running out from that side will allow, and once past these making all haste to the left to clear the breakers below. In high water the rapids can be avoided by making a portage of about half a mile along the river's right bank.*

We could see that according to McConnell the upcoming run through Little Canyon would be a challenging one.

Shortly before dusk we grabbed the canoes, ferried across the back channel and pulled them up on the shore to tie them to willows. It took only minutes to get soaked again, as we struggled through the thick tangle of dripping willows and then climbed the steep, rockstrewn slope of the high bank for a better look at the rapids.

Below the island, from where the smoke of our cooking fire curled up through the spruce trees, the Liard formed a tight S-turn before rushing into the canyon itself. The rocky slope grew steeper and more slippery the closer we climbed to the top.

But it was worth the time-consuming trek. Little Canyon, we could see at once, had changed little since the passing of the voyageurs and McConnell. His report was still accurate. To the left a narrow but clear chute offered a safe passage into the eddy below the inside corner of the bend and was indeed the only route to take. It was just out of reach of the heavy breakers that even now, in low water, rolled more than halfway into the churning waters from the right shore.

But to see what lay beyond the first eddy and to get a glimpse of the whirlpools which McConnell had also mentioned, at the end of the canyon, we had to climb higher. The gushing river roaring below, we scrambled across a sheer rocky ridge, hanging on to a few rose bushes that sometimes tore loose and always left scratches on hands and arms. Behind the ridge a short slope dipped down to a large flat terrace, covered with tall, yellow, rangy grass as thick as a hay crop. We were now standing on top of a rocky overhang that dropped twenty metres straight down to the river where it fought its way through the narrowest part of the gorge. From our grassy perch we looked straight down into the forever shifting eddies and whirlpools. Each one seemed large enough to suck up an entire canoe if it happened to get into the wrong spot. We would have to stay outside of these churning, changing eddies, to anticipate the river and paddle hard to get through safely. Farther downstream, where the river bent out of sight, was another large eddy. It would be our goal.

Satisfied that at the present water level we could safely run this entire stretch, we climbed back down to the canoes and then slipped across the now dark channel to the island.

It was fully dark now. Large chunks of dry logs flared up the fire. Standing by the fire, shifting and turning, we dried out, while a pot of tea sweetened with honey and a generous shot of "medicinal" rum relaxed mind and body. There wasn't much talk, we were all tired from the day's work and trying to memorize the safest approach to the first real rapids that lay ahead. Then we crawled into the sleeping bags and fell asleep to the steady roar of the rapids, to the wind that lashed the trees and to the rain-drops that fell hard as hail and occasionally drummed onto the tight nylon fly of the tent beneath the spruce trees.

It was still pouring rain the next morning. We coaxed the fire into a hot blaze and soon enjoyed the rich brew of fresh strong coffee.

You could tell that there was whitewater ahead. Everybody was more tense than usual and the morning jokes were flat and forced, and few and far between. Obviously I was not the only one with butterflies in my gut. Wild rapids on a strange river are something which not even an expert canoeist attacks with nonchalance. And I was far from being an expert! In silence we loaded the canoes, lashed the loads securely and then fastened the canvas spray covers. We had designed these covers of waterproof canvas ourselves the previous winter when we had also built our own cedar-strip canoes in Harvey's workshop. They covered the entire canoe and were fastened to the underside of the gunnels with press buttons. They were an essential safety factor in running whitewater with a loaded canoe. If we missed the ideal route through Little Canyon even by a metre and were driven into the tail-end of the large breakers, the water would surge over the gunnels. The spray covers held the water out of the canoe thus kept us and the load dry.

Harvey and his brother Doug took the lead. Quickly and without difficulty they shot through the narrow chute between the first rocks, charged across the high waves and whipped into the eddy below the overhang we had stood on the previous evening. One by one the other canoes followed. So far, so good. We all relaxed and began to enjoy the challenge.

"OK, we'll dodge these whirlpools and then land in that eddy behind the rocks over there on the right side," Harvey said. "We will tie up the canoes and then climb back upstream over those rocks," he added, pointing to a series of rocks jutting far into the river from the right hand shore. "From there we should be able to get a few good shots when you guys come through the standing waves and then turn in below us." Then, packing our cameras into a watertight bag, he and Doug pushed off.

It took almost ten minutes for them to reappear. I was beginning to worry, but the drenched rocks were slippery and walking was slow. Then, crouching on the rocks and cameras at the ready, Harvey raised his arm and gave the signal.

We were off. Wink and Jeff were in the lead now and Frank and I at the rear. A quick peel-off brought us into the main current.

Holding tight into the next chute and backpaddling to keep with the speed of the current, so as not to swamp in the high waves, we shot past Harvey and Doug, who frantically tried to snap pictures with three cameras at once. Then we were approaching the whirlpools. We tried to read their patterns as they surged across the channel, filling up only to emerge again, unexpectedly, in roaring boils, to open up yet again, suddenly, into churning vortexes. Adrenaline pumping freely, we had no time to be scared. We paddled hard and whipped behind the large boulder, to pull into shore where Harvey's canoe danced on the water. Exulted now we were ready to tackle more of the same.

It had gone well. Only once did Frank and I have too much forward speed, when, instantly, a high wave crashed into my lap. Luckily the shroud held off most of the water.

As suddenly as it had started, the rain stopped. Sunlight now streaked through the grey clouds and dropped onto our wet faces and sparkling foliage of the willows like a drizzle of gold. We took off the rain slickers and enjoyed the warmth of the sun breaking through the last of the clouds. Past Little Canyon the river widened and flowed fast but even between low hills. A light breeze fanned over faces and bare arms. Good weather seemed well on the way.

Before us, stretching over a distance of perhaps a hundred metres, lay another tight spot where the Liard was confined between rocky walls scarcely fifty metres apart. Still, it did not look too dangerous. Above and below the restriction were large bays where the river spread out over two hundred metres and more. In the gate itself though, the stream had to be very deep and the usual whirlpools marked the exit of the short canyon.

We were watchful as we drifted closer and then parked in the last eddy upstream. Harvey and Doug worked their way through and pulled onto the shore below the gap. They had no problems. Now it was Wink's turn. A straight forward paddle, a slight shift to the right and he too was out in the open again. Now Adolf pushed off. Frank and I were close behind. Suddenly, as if grabbed by an invisible force, Adolf's canoe tore ninety degrees around and only an instinctive low brace prevented an upset. His canoe then shot toward the sheer, jagged rocks on the left bank. A crash seemed unavoidable. Frantically Adolf tried to turn the canoe, while Dale in the bow first backpaddled and then leaned far out over the gunnel to draw. Nothing had any effect. The canoe still headed straight for the rocky wall, carried by a strong current. Only as the bow was but a metre from certain destruction, and Dale was trying to scramble backwards out of his seat, another current grabbed the canoe and shot it back into the main stream alongside the rocky wall. Once again, only instant bracing kept them upright.

I watched their predicament with utter amazement, and out of the corner of my eyes, and we too were now fighting the current to slip into the eddy below. Just why we did not hit the same cross-current—we had been right on Adolf's tail—is astonishing.

The river stayed swift over the next twenty-five kilometres. Three more times we manoeuvered through tight gates, each time doing the "Liard-Waltz," as we had dubbed the constantly shifting eddies and whirlpools and boils. Between the canyons though, the Liard opened up to the most beautiful vistas. On the sandy beach of an island we stopped. This point of land was packed with driftwood. Grateful for the ready-made supply, we built a fire and stretched out in the warm sand to wait for the soup to boil. The afternoon still lay before us, and with it the beginning of another challenge: Cranberry Portage.

As I recollect these events, something that happened on a later trip comes to mind. Four years later, when some of us again came through Little Canyon on yet another trip down the Liard River, our experiences were much more dramatic and dangerous. A cold, wet summer in the North and a heavy storm three weeks before our arrival in the country, had kept the Liard unusually high for early September. We feared the worst when we approached Little Canyon. We knew that the gorge could become a real obstacle with the water three or four metres higher than normal.

Apprehensive, we stopped on the island to boil a pot of tea. Nobody had been there since our last passing, nothing had been disturbed. The campsite in the tall spruce trees was exactly as we had left it, the tent poles still leaning against the tree where we had put them. People, that was certain, did not yet stumble over one another along this stretch of the river.

Our concerns about the Little Canyon were justified. We again climbed the rocky slope to the top of the overhang that overlooked the short canyon and the giant whirlpools beyond. But this time the river ran wild below us. Gone was the nice clear chute to the first eddy that we had used to enter the canyon. In its place were half a dozen strong diagonal waves, crashing heavily into the vertical rock. The breakers from the other side of the river spilled all the way across the stream, only to clash with the diagonals in the middle of the main current. Gone too, was the large, almost placid eddy where we had gathered the last time, halfway into the rapids. The rocks from where Harvey and Doug had taken the photographs, were now buried beneath crashing waves; rooster tails shot three metres into the air. In the place of the eddy was now a rushing, gurgling, madly spinning pinwheel of a whirlpool, just waiting, it seemed, to swallow a careless river runner. Farther downstream, where the river bent out of sight, were more strong eddies and more high waves.

Standing on the grassy bench high above the seething waters—this time basking in the warm fall sunshine—we wondered what to do. There was, we could see, a definite route through all this madness, but a single canoe did not have much of a chance of staying upright in this heavy stuff; the breakers would soon swamp it. But here we had a chance to try out an idea we had discussed many times when talk turned to what we would do if somebody got hurt on one of these trips and we would have to come through this heavy stuff in a hurry. After many discussions, we had our answer. We would tie two canoes together with strong poles. By rafting up, as we called it, we would not have to worry about staying upright, and at the same time did not have to resort to a time-consuming, laborious portage over difficult terrain.

We climbed down from the overhang and pulled the canoes up on the soft sand above the rapids. Cutting poles and then lashing them across the canoes at the three thwarts, leaving a half-metre space between the two crafts for the water to surge up, we tightened down the spray covers and were ready to go.

As we pushed off and headed into the main current, we were surprised at the ease with which we could handle the two rafted canoes. We cut across the diagonals, peeled off and spun into the churning eddy below.

Again we peeled off. Then suddenly, without warning, it happened. Below the stern of the righthand canoe where I was sitting, a giant hole yawned open. The whirlpool, coming in from behind and under the canoe, stopped all forward motion. Slowly at first, and then faster, amidst the sucking and gurgling noise of the vortex, the canoes began to slide backwards into its churning maw. Almost instantly the water surged over the cover and up to my chest while the bowman in the other canoe hung suspended in the air, a metre-and-a-half above the water. The poles holding the crazily tilted canoes together, screeched and bent, threatening to break apart. For what seemed an eternity we sat motionless. Finally Harvey's yell, "Paddle! Paddle!" snapped us into action. Adrenaline flooded our veins as fast as the river's current. Strength, doubled with fear, returned. Straining we dug in and pulled on the paddles.

Slowly the canoes pulled loose. The whirlpool shifted, the vortex filled up and the canoes shot out of it. But the Liard was not yet prepared to let us go. Another large eddy came shifting in from the right. It spun us around and spun us around again, while the water cascaded over the bows of the canoes until, finally, we regained control. Putting all our strength into a few hard strokes, we ferried across the last chute, cut across some heavy diagonals and whipped into the eddy behind them.

Only when safe ashore, did the full realization of our close escape hit us. Taut nerves erupted into hysterical laughter. Later, relaxing over mugs of steaming tea, we all agreed—rafting up was the only way to out-manoeuvre such dangerous rapids and whirlpools. Solo canoes were defying immeasurable odds in such places and would have to be portaged.

CHAPTER 3

Cranberry Portage

You can hardly conceive the intense horror the men have to go up to Frances Lake. (Headwater of the Liard River). They invariably on rehiring endeavour to be exempted from the West Branch (Liard). The number of deaths which have occurred there is fourteen, viz. three in connection with Dease Lake and eleven in connection with Frances Lake and Pelly Banks, of these last three died from starvation and eight from drowning.

—HBC clerk, (1853)

AFTER LUNCH ON THAT THIRD DAY into the first trip we drifted down the right-hand bank, approaching Cranberry Rapids. We were again apprehensive and nervous, the words of McConnell's report we carried with us were in all our minds:

The rough water of Cranberry Rapids and Portage has a total length of a mile and a half, but there is a stretch of comparitively undisturbed water about half way down. The upper part of the rapid is exceedingly wild, as the bed of the river is filled with huge angular masses of rocks against which the current breaks with frightening speed and violence. No part of the channel is clear and a glance at the forbidden array of foaming breakers and whirling eddies showed at once the utter hopelessness of any attempt to run it with our heavily laden boat. We passed it by portaging along the right-hand bank.

The Alaska Highway Bridge at the Liard River near Liard Crossing.

Remembering these words as we approached the rapid did nothing to ease the tension we felt. As a fellow canoeist once said, "You should not read *Deliverance* and then go canoeing."

We had studied the Cranberry Rapids when we drove up the Alaska Highway to our starting point of the trip at Watson Lake. Along Cranberry the highway runs close to the river. We climbed down the densely wooded slope for a firsthand look from solid ground.

Obviously the rapids had not changed with the passing of the years since McConnell's trip. Large waves crashed against dark, dripping, shaly rocks—their edges razor sharp. Rocky ledges dropped off, sometimes a metre-and-a-half high and were equally sharp. Crisscrossing currents, interspersed with rooster tails, shooting spray three metres and more into the air, did not leave any clear chutes. All together, rocks and wild water indicated rather strongly that this, the highway side, would be a good place to avoid.

Now we were coming down the other side of the river, where McConnell had portaged his gear and supplies over the old voyageur trail. On the opposite shore, close to the bank, a rocky island jutted into view, with its high crown of driftwood, a three-metre tangle of bleached logs deposited there by heavy flood waters. We had picked this island as our landmark to the approach of the Cranberry Rapids. Directly behind the island, and confirming that it was indeed the right one, the Alaska Highway swooped elegantly down a steep hill, almost touching the water's edge, and disappeared again into a thick stand of trees, before climbing another hill.

Already the current flowed with tremendous speed. And even though the rapids were still out of sight behind the next river bend, their roar assaulted our ears like the thunder of a distant diesel train.

It was time to slow down and head for shore. We backpaddled and edged closer to the bank, looking for a place to land. There was none. The rocky walls grew steeper and higher and were now falling straight into the rushing river. One question kept nagging my mind, had we waited too long already with our move to shore? We backpaddled more strongly to slow the canoes down and still the rocky shore rushed by at great speed. A shaly outcrop stretching ten metres or more into the river and throwing off a series of heavy rolling diagonal waves came up much too fast. From this rocky spit the current gushed toward the centre of the stream where the first white breakers of the rapids were now only a hundred metres away. But behind the black, glistening rocky tongue lay our chance. A large eddy had formed there—actually more of a whirlpool—with the current along the shore flowing upstream just as fast and furious as it flowed downstream in the main channel.

There was no time to waste. We cut across the diagonals, as close to the rocks as we dared and whipped into the eddy, our canoes almost on top of one another. Throwing back the spray covers, the bowmen were ready. Bowline in hand they jumped into the hip-deep swirling water, leaped to shore and tied the canoes to the trunk of a small, lone spruce tree whose roots had somehow managed to take hold in an earth-filled crack of the rocky wall.

The narrow bay behind the rocky tongue where we were now parked was hardly wide enough to hold two canoes side by side, let alone four. Beside it the river ran with frightening speed. On the shore side the slope was almost vertical, its rocks covered with a thin layer of moss. It was more than ten metres high—to climb it, as we knew we may have to, with canoes and gear, seemed almost impossible.

But we were safe for the moment, uneasy perhaps, but safe.

Downstream, less than a hundred metres away, the Cranberry Rapids were noisy, wild and spectacular. Black boulders the size of small houses split the river into a number of channels, each, it seemed from our eddy, more formidable than the last. Between the boulders were sharp ledges, now and then revealing their knife-like edges and running at an angle to the current. They lurked there ready to slash everything that was foolish enough to come their way. Behind the boulders, wet with spray, the water churned and boiled in great souse holes and giant keepers—vertical eddies that held everything fast that got into them. We agreed with McConnell: safe passage was indeed impossible.

There was only one alternative. Canoes, gear and supplies for eight men for one month, had to come up the steep, rocky bank, to be carried over the portage trail to the foot of the rapids. Frank, although a good paddler, was not yet used to the the wildness of this Canadian river, and was pessimistic.

"It's impossible, it's impossible," he kept mumbling over and over in his native tongue, shaking his head. "You'll never get the canoes up this wall." But he also knew that there was no other way. A quick climb to the top of the cliff and a short walk downstream high above the water, had only confirmed that we had indeed landed in the very last eddy above the rapids.

Our friends, who made up the rest of the group, did not seem all that worried with the prospect of a tough carry. And the matter-of-factness with which they tackled this first portage left no doubt of the outcome. Encouraged, I grinned at Frank and said, "Not to worry, you are in good hands."

Dale grabbed a coil of rope from under the cover of his canoe and strung a line from spruce to spruce to the top of the bank. Now we had something to hang on to as we started packing supplies and loose gear pack by pack up the steep, rocky slope, to deposit the loads in a large pile on a carpet of moss on the first level bench high above the river. Harvey and Wink, up to their hips in the cold, swirling water, unloaded the canoes. The rest of us struggled under the loads and with the treacherous footing.

Finally there were only the canoes left at water level. The slope was too steep and too rough for two men to lift the canoes together, but everybody could pull on a rope. Shivering from the long immersion in the cold water, Harvey climbed onto dry land, reached down and threw the first of the canoes onto his back, the centre thwart resting squarely on the three topmost bones of his spine. Stumbling, and then catching his balance, he grunted from under his load, "O.K., guys, pull!" And pull we did on the rope we had fastened to the bowline. And up the cliff came the canoes, one by one, until at the end an exhausted but happy crew lay in the deep moss beside the pile of equipment and supplies.

Still before us was the portage itself. "But what the hell," we thought, "after this anything is child's play," hoping we could locate the old trail. Walking was not easy in the thick carpet of moss and the tangle of low-bush cranberries that had given the rapids their name. We axed our way through the heavy growth of willows and alders, poplars and young spruce trees, and moved farther inland to search for the old trail. We also had to avoid a couple of steep gullies that plunged down to the water. Where the ground finally levelled out, high up on a second bench, Wink found the trail. He smiled when he pointed to the scar of an old axe blaze that marked an ancient spruce tree. Thirty metres away was another blaze, this time hewn into the bark of an old cottonwood tree and farther on there was another and another. Underfoot was the trail. Trod centuries ago by the Indians, then by voyageurs, explorers and Klondikers, in some places it was worn twenty or thirty centimetres deep and overgrown with moss. Looking along it we could clearly make out the right-of-way of the portage track. It ran in an almost straight line, paralleling the river and followed, as far as we could make out, the level ground some distance from the river. It seemed if we listened carefully, we could hear the echo of these early traveller's voices ringing in the cool air.

Along this trail the tough but always cheerful—if one could believe the old journals—voyageurs had trotted, bent double under their standard loads of two pieces, each weighing forty-five kilograms or ninety pounds. Along this trail they had struggled under the weight of their canoes, and along this trail they had cursed the steep banks, slippery after a hard rain, the hordes of mosquitoes and other stinging devils against which the bear fat they had smeared on their faces had little effect.

Along this trail, pale and weary and only too often starving, ridden with scurvy, hollow-eyed and helping their friends, had struggled the Klondikers. Burdened with heavy loads or pulling sleds and toboggans, they had spent weeks in the icy cold of midwinter. For many it was now almost a year since they had left the relative comforts of Edmonton, or Hamilton, or Chicago. Their supplies were running out and all of them wondered just how long this insane struggle was to last. Forgotten were the dreams of rich gold finds. Survival was now the only concern.

J. G. MacGregor writes in his book *The Klondike Rush Through Edmonton*:

> For those Klondikers who took the water route Fort Simpson, some 1,700 kilometres from Edmonton, was nearly the halfway point in their 4,000 kilometre trip. It was also the place where the first alternative route presented itself. That route lay up the Liard River, and several parties of Klondikers decided to follow it.

Unfortunately, most of these men had apparently never heard what the Hudson's Bay voyageurs had to say about this river, nor had they read the reports of McConnell and Dr. Dawson, published by the Geological and Natural History Survey of Canada only ten years before.

They had not read McConnell's letter to a friend:

> *Between Little Canyon and forty miles below the Devil's Portage—a distance of about a hundred miles—the Liard falls over 1,000 feet, and as forty miles of this is good water, you can imagine the state of the rest. It would be bad enough if it had room to flow, but it is penned in by canyons, often less than 150 feet wide, every few miles, and then whirls and boils in an incredible manner. We worked our way by sheer muscle carrying boat and stuff through forest and over high hills. My men turned out well and worked without a grumble. As for myself, I have hardly a stitch of clothes left, as they are torn to pieces and left hanging on the brambles and roots along the portages.*

From the pile of packboards, and food, and paddles, and lifejackets, and canoes, we gathered the first light load for a walk to the other end of the portage. Taking turns with the two axes we carried, we cleaned out the trail until it was again wide enough for the canoes. It was not easy and sweat flowed freely but we hacked and broke and pushed our way through the junglelike growth. Windfall was the worst enemy. Often we were forced to crawl under the downed trees, or climb over them, when their trunks were too heavy to be cut with our light axes. We wished for chainsaws.

For a distance of about 800 metres, the old track followed level ground. But when the bench started to angle down toward the river again where it ended in a sheer drop—the roar of the river muffled by the dense growth of trees along its banks—we were forced to climb yet another sidehill and to push through yet another gully.

Halfway along the trail we dropped the loads and fought our way down the slope to the steep banks of the river. Thick underbrush and arm-thick willows, straight and dense like bamboo, barred the way. Only when we reached the very edge could we see the boiling waters of the Liard, and looking at it we were glad to be on solid ground.

We had come to the edge of a large side channel. Dark slabs of shale dropped straight down more then ten metres into the channel of the same width. Separated from the main current by a narrow, solid slab of rock, the water had cut a turbulent path for itself. The sharp-edged rocks at the bottom end of the channel did nothing to impede the frothing waters. We could only imagine what this section would look like when the river was flooding. McConnell had had a taste of that. We struggled back to our loads, satisfied to walk a little farther.

The trail was swinging toward the river again and after dropping over a steep bank, and bypassing a stand of giant spruce trees, one with a metre high old blaze, it led to a slightly sloped beach strewn with boulders and stones. In front of the beach the river surged in heavy, metre high waves. We scouted ahead and pushed our way across the tip of a narrow peninsula with heavy undergrowth, to finally step out of the bush and onto a beach with soft, white sand.

An hour had passed since we had started out. It was getting late but there were two more trips to be made. Each time we dumped our loads onto the sand alongside a large, polished driftwood log.

Frank and Jeff had not made the third trip. Instead they had set up the kitchen and built a fire to boil soup. We wolfed it down. It hardly cut the edge of the hunger. "There is nothing like fresh air and exercise to make you hungry," grinned Harvey.

Daylight was not going to last much longer and there remained still one more trip for the canoes. While the rest of the crew set up the tent, cooked supper and then baked the daily bannock,

Wink, Adolf, Harvey and I trudged weary limbs over the trail once more.

It was now nine p.m. and I was getting tired. It had been a busy day and a long one. But the worst was yet to come, the grunting and sweating under the heavy load of the canoe, stumbling in the approaching darkness over roots and windfall, always looking for the ideal tree with a handy fork to wedge the bow against, to rest a tired back. The topmost bones of the spine took a beating that night. Not yet used to the load, mine were nearly rubbed raw in my struggle down the trail, from tree to tree, my legs shuffling along as if they belonged to some old man, all thoughts focused on the pain in my neck, my breath coming in short gasps.

Soon the others pulled away. I was alone and lost track of time. Finally there was the last steep slope. The stern was dragging on the ground, my legs half bent, my back doubled over in an effort to find a virgin spot for the unrelenting pain. Then there was the boulder-strewn beach and beyond it, my feet sank into the soft sand. And down came the canoe, safe into Harvey's hands.

I sat on a log. The crushing weight of physical exhaustion intermingled with the mental buoyancy of hard-earned success. It was done, the first portage was behind us. After a quick dip into the Liard's icy water, after a leisurely supper and endless cups of sweet tea, the last fortified with a generous shot of rum, and after a helping of freshly baked bannock, buttered thick and heaping with sweet jam, we knew, that from now on the trip would be all right.

This outfit had been shaken down.

The next morning, long before the sun topped the higher hills, we were up to our backside in the cold, swirling water again. The sandy beach was deceiving; if we had thought that reloading the canoes would be an easy task, we were wrong. The strip of sand did not run into the river gradually but ended abruptly in jagged rocks, dropping off very quickly into deep, surging water. To make matters worse, the river still had not settled down after its rampage over Cranberry Rapids. In fact, it was already gathering momentum for yet another battle with the rocks, as if it wanted to tear them loose. That was bad medicine for our canoes. Each pounding wave lifted them, tried to tear them out of our hands and then threatened to break them up as the water receded in a noisy, gurgling surge, only to rush back for another attack a few seconds later. There was nothing to do but to wade out into the cold deep, hang on with difficulties in the strong current, and hold the canoes off the rocks, while one's partner stowed away the boxes and packs and then tied down the spray cover.

CHAPTER 5

The Voyages of John M. McLeod and Robert Campbell

We now returned down stream to Fort Halkett, which we reached about the middle of Sept. with our canoes loaded with provisions. We saw no Indians, nor trace of them during the entire trip. Our arrival was very opportune as next day the outfit and packet arrived from Fort Simpson in charge of J.B. Bruce, the well-known guide.

Early in the morning of 18th Sept. Bruce, Mr. Mowat and 7 men left for Fort Simpson and before noon that day, as I learned with profound sorrow three days afterwards, Mr. Mowat and 5 others drowned. While running a rapid, their canoe went to pieces and six sank to rise no more. Bruce and 2 others managed to swim to shore, and came back to tell us the mournful news, which was inexpressibly sad, as I knew them all so well.

—Robert Campbell's Journals (1808-1853)

THE DIFFICULTIES WE EXPERIENCED in loading the canoes the morning after portaging Cranberry Rapids, the constant exposure to cold and raging waters, was just the kind of thing the early explorers had already cursed, and then accepted as being all in a day's work on the river. Men like John M. McLeod and Robert Campbell and their engagés, the voyageurs, the pioneers of the Liard, the men of the fur companies, they all knew the fierce bite of the cold waters.

What follows here is a brief portrait of the men who explored and opened up this waterway in search of furs for the Hudson's Bay Company. Knowing the history of a river, any river, when you canoe its course adds immeasurably to the enjoyment of such a trip. It also humbles any modern day voyageur for no matter how tough and experienced that modern day canoeist may be, the men who came before him, exploring and looking to expand the fur trade, knew hardship beyond any imagination.

George Mercier Dawson, the well-known geologist—himself well versed in river travel—in his *Report on an Exploration in the Yukon District, N.W.T. and Adjacent Portions of British Columbia, Geological Survey of Canada*, 1887-88, had nothing but admiration for these men when he wrote:

> *The utmost credit must be accorded to the pioneers of the Hudson's Bay Company for the enterprise displayed by them in carrying their trade into the Yukon basin in the face of difficulties so great and at such an immense distance from their base of supplies. To explorations of this kind performed in the service of commerce, unostentatiously and as a matter of simple duty by such men as Mackenzie, Fraser, Thompson and Campbell, we owe the discovery of our great Northwest country.... Less resolute men would scarcely have entertained the idea of utilizing, as an avenue of trade, a river so perilous of navigation as the Liard proved to be when explored. So long, however, as it appeared to be the most practicable route to the country beyond the mountains its abandonment was not even contemplated. Neither distance nor danger appeared to have been taken into account and in spite of every obstacle a way was found and a series of posts established, extending from Fort Simpson on the Mackenzie to Fort Yukon....*

Just how far removed these posts were is also indicated in a report which James Anderson, Chief Factor of the Hudson's Bay Mackenzie District, wrote in 1852:

> *At the time of the establishment of Fort Yukon and Fort Selkirk, and for many years afterwards the returns from these furthest stations reached the market only after seven years, the course of the trade being as follows:* Goods. *1st year reach York Factory; 2nd year Norway House; 3rd year Peel River and were hauled during the winter across the mountains to La Pierre House; 4th year reach Fort Yukon.* Returns. *5th year reach La Pierre House and were hauled across to Peel River; 6th year reach depot at Fort Simpson; 7th year reach market.*

John M. McLeod's name in connection with the Liard River comes up in a letter he wrote to John McLeod senior in 1831. John McLeod senior had established Fort Halkett at the junction of Smith River and Liard two years earlier in 1829, and John M. McLeod was appointed Chief Trader to that post in 1831. In June of that year he confirmed his readiness for the trip up the Liard in a letter to his namesake:

> *I am to take my departure in a couple of days on a surveying voyage by the West Branch of the River au Liard, a stream of as yet unknown (only from Indian reports) to any of us in this quarter, we expect to be able to cross to the West of the Mountains by the present intended route which if found practicable will soon augment the returns of this quarter, how the voyage may terminate time can only determine.*
>
> —Letter from John M. McLeod to McLeod senior, (24th June 1831).

You can get an idea of how things went on that trip from

another letter which John M. McLeod wrote to his superior after he got safely back to Fort Simpson:

> My voyage in exploring the West Branch terminated to the utmost satisfaction, having succeeded in navigating that unknown stream to its source, a distance from its junction with the Rivière au Liard of upwards to 550 miles. The navigation of the river is both difficult and dangerous, arising principally from the perpendicular and stupendous banks of the river in many parts, as projecting points of which form some dangerous Cascades and Rapids, in running of which when on our homeward voyage, our canoe broke in three pieces and exclusive of all our property unfortunately, two of my crew met with a watery grave, we were at the time upwards of 300 miles from any establishment wether provisions or craft and thus I experienced that necessity is the mother of all inventions. Not even an axe did we save, and altho not without time and trouble, with only a single knife, we succeeded in rebuilding our canoe to nearly its original size, and sufficiently seaworthy to float us down the stream which brought us safely to Fort Simpson on 10th September (1832).
>
> —Letter from John M. McLeod to John McLeod, senior, Fort Simpson, 16th March, 1833.

Robert Campbell first saw the Liard a year and a half later, when he arrived in Fort Simpson, on October 16th, 1834, from Fort Chipewyan. Dr. Dawson gives Campbell most of the credit for establishing the fur trading posts along the Liard and beyond. He wrote:

> After the abandonment of Dease Lake post in 1839, Mr. Campbell was, in the spring of 1840, commissioned by Sir George Simpson to explore the North Branch of the Liard to its source, and to cross the height of land in search of any river flowing

westward, especially the headwaters of the Colville, the mouth of which on the Arctic Ocean has recently been discovered by Messrs. Dease and Simpson.

From this and other sources, among them Robert Campbell's own journals, covering the years 1808 to 1853, which at one time were thought to have been lost, the following sketch of this remarkable wilderness traveller and explorer emerges.

Robert Campbell, a tall and powerful Scot from Pertshire, joined the Hudson's Bay Company in 1830 when he was twenty-two years old. He came to the Mackenzie District four years later and was put in charge of the post at Fort Liard, some 350 kilometres upstream from Fort Simpson on the Mackenzie. Campbell then spent the next ten or twelve years exploring the river between Fort Simpson and Dease Lake, and later went up to the headwaters at Frances Lake and finally over the mountains to the western slopes.

In 1837, for example, he was on his way to Fort Halkett at the mouth of Smith River. The previous summer a party of voyageurs under the leadership of a Mr. Hutchinson, who had wintered at Fort Halkett as the successor of John M. McLeod, had been ordered to go up the Liard and establish a post on Dease Lake. Upon reaching Portage Brulé Rapids the entire party had panicked when a rumour came up that hundreds of Russian Indians were advancing on their camp to murder them all. Campbell writes:

> *A panic sized the whole party and they ran down the bank pellmell, jumped into their canoe and off downstream, never halting till they reached Fort Liard where they stopped for a few days and then came down to Fort Simpson.... During the winter I volunteered to go to Fort Halkett.... As a preliminary step in the undertaking, I left the depot in March with a party of men for Fort Liard, where canoes for the trip had to be made and birchbark for that purpose obtained. We had the canoes*

> *finished before open water and all I wanted was canoemen and Hunters. We found it very hard to get men, either whites or Indians, willing to go. The panic of the previous year seemed to have spread all over the district.*
>
> *However about the middle of May, I managed to muster a sufficient number of men, say 2 crews or 16 in all, and started . . . .*

Campbell, however did not have much luck that early spring. Three days into the trip half the men cleared out during the night and the rest refused to go any farther. Campbell was forced to return to Fort Liard and get refitted. A few days later he managed to get new crews and started out again. It was now getting late in the season and the summer freshet—as Campbell called the spring run-off—came on before the party reached Hell's Gate and the Grand Canyon of the Liard. Campbell writes:

> *When the river is in flood no boat that is built could ascend from the Devil's Portage, as the current is not only strong but is full of rapids and whirlpools and rushes between perpendicular walls of rocks 2 or 3 hundred feet high. Our progress was necessarily very slow and I was kept in constant anxiety lest the men would lose heart, as the croak of a frog or the screech of a night owl was immediately taken for an enemy's signal. These imaginary dangers gave me more trouble and concern than the real ones . . . .*

Campbell and his crews finally did reach Devil's Portage at the upstream end of the Grand Canyon of the Liard and in three days of hard work got canoes and cargo over the portage. Campbell then decided to winter at Fort Halkett because it was now too late in the season to go on and because some of the Indians he had with him had to return to Fort Liard. He then writes:

> *However, I was determined to go to Portage Brulé, the spot Mr. Hutchinson and Party had evacuated so hurriedly the year before, and which, on reaching, we found just as they had left it. The goods were scattered all the way down to the water's edge, just as they had been dropped by the men running to the canoes. Of course, everything was spoilt, except such articles as ball, shot and the provisions eaten by wild animals. . . . As prearranged the trading outfit was forwarded to us from Ft. Simpson to the Devil's Portage in Sept. The winter passed uneventfully in our quiet retreat in the heart of the Rocky Mountains . . . .*

Campbell's work came to the attention of Governor Simpson; it brought him a promotion from postmaster to clerk with an advance in salary. In a letter to Campbell, George Simpson wrote in July 1837:

> *Rest assured that merit will in this service always meet its reward. Let me beg you that your attention be particularly directed to pushing the trade across the Mountains and down the Pelly River . . . . and Robert Campbell is not the man I take him to be unless in due time he plants the* HB *Standard on the shores of the Pacific.*

In the following spring a party came up from Fort Simpson with extra hands for the intended summer's work and Campbell started for Dease Lake with two canoes each manned by eight men. Again the water was very high and the men had a tough time of it, but reached Dease Lake in July of 1838. While some of the men started the construction of the new fort, he and his interpreter, Francis Hoole, and two Indians—Lapie and Kitza (they were to be his companions on many adventures later on), started out to explore the western face of the mountains. During these explorations Campbell met with a great number of Indians in the valley of the Stikine River, among them a certain Chief Shakes who acted as the agent of the Russians who at the time held Fort Highfield at the mouth of the Stikine River. Later he also met a chieftainess of the Nahanies, a "fine looking woman, rather above middle height and about 35 years old"—a woman who was to save his life more than once later on.

Campbell then left Dease Lake with the Indian Lapie to get the news of his discoveries to Fort Simpson and also to get the necessary supplies and a trading outfit for the new post. The two men used a pine bark canoe as far as Fort Halkett and there changed to a much more reliable birchbark craft. They reached Fort Simpson in mid-August after "nearly sinking at Hell's Gate rapids." Here is what Campbell himself had to say about that trip:

> *It was by no means an ordinary undertaking for us to attempt to run down the strong and swift current of a river so full of rapids, whirlpools and other dangerous places, in a canoe of the frailest material such as ours was. We trusted our guns for food. At Fort Halkett we changed our pine bark for a small birch canoe we had left there in the spring. We then resumed our route reaching Devil's Portage the same day . . . The next day we were passing between the rocky cliffs and through the rapids leading for miles to Hell's Gate. Our canoe being so small, the waves kept dashing over it and we shipped so much water that we had constantly to bail her out or empty her whenever we could get a ledge or a strip of beach to stand on. When some distance above Hell's Gate and enclosed on both sides by high perpendicular walls of rock, with not a spot where we could land, our canoe sprung a leak. We thought we were lost . . . in the midst of our anxiety we saw a pine tree growing on a ledge near the water's edge. We made for this, succeeded in making a landing and to our great relief found pitch on the tree. Lighting a small fire with flint,*

steel and touchwood, we turned up the canoe and pitched it where required . . . and about 20 August arrived at Fort Simpson.

Although Robert Campbell went there at the risk of his life, Mr. McPherson who ran the post there, "was deaf to all my [Campbell's] entreaties for a few extra and much wanted supplies," and Campbell left for Fort Liard with a very limited outfit. That same fall he proved for the first time what an incredible traveller he was. He met his own men at the Devil's Portage and then went upstream again to spend a miserable winter at Dease Lake. It was the first of many miserable winters he spent in that country but this was a particularly bad one. All game seemed to have left the country entirely, and Campbell and his men were at the verge of starvation on many occasions. When things were at their worst he was forced to send some of the men back to the Liard to fend for themselves. The rest of the party was spread out in several camps along the lake shore. Luckily the Nahanie chieftainess, whom Campbell had met earlier, came by with a hunting party and saved their lives by sharing food with them.

The following spring, May 1839, Campbell and his men descended the Liard as far as Fort Halkett where he had been put in charge. And here, in September of that year, he received another letter from Governor Simpson which explained that the Hudson's Bay Company had leased the Russian territory for ten years. Simpson wrote that this arrangement made it now unnecessary to extend the company's operations from the east side of the mountains or Mackenzie River, "as we can settle that country from the Pacific with greater facility and at less expense." Instead he was to extend his explorations to the headwaters of the Liard.

And so in May of 1840 Campbell set out up the Liard again. This time they bypassed the mouth of the Dease River and continued until they reached the source of the river, a beautiful lake in the shape of a spur. This was named Frances Lake, in honour of the governor's wife. He then pushed on. First he canoed up the western arm of the lake and then, on foot, up a river valley leading northwest. He crossed the watershed and six days out from Frances Lake saw from a high bank a large river in the distance. It was flowing northwest.

"I named the bank from which I caught the first glimpse of the river Pelly Bank," wrote Campbell, "and the river, Pelly River after our home governor, Sir H. Pelly."

Campbell and his men then returned downstream to Fort Halkett only to once again spend the winter near starvation. In the spring of 1841 he returned to Fort Simpson with the news of his discoveries. Travelling the difficult Liard now became almost routine.

In August of 1842, Campbell with an outfit of two large canoes, came again up to Frances Lake. This time it was to build a trading post there and to spend the winter. Using Fort Frances as his headquarters, he then undertook various explorations over the next few years; explorations that eventually led him to the discovery of the Lewes and Stewart Rivers. The explorer and his party of Indians and voyageurs then travelled to La Pierre House on the Porcupine River, crossed on foot over the mountains to the Peel River and finally returned to Fort Simpson by coming up the Mackenzie and thus completed a full circle by an entirely new route.

On August 21, 1852, Fort Selkirk, built at the junction of the Pelly and Lewes Rivers, was overrun by a party of Chilkats—coast Indians from Lynn Canal—and Campbell and his men were lucky to escape with their lives. They arrived at Fort Simpson with the news of the Chilkats late in October amidst chunks of drifting ice. This was, as it turned out, the last time for Robert Campbell to use the shortcut of the Pelly River and Frances Lake on the Liard route. The reason for this became clear later on. At the moment Campbell still had only one thing on his mind—to get a new outfit for Fort Selkirk, to get permission to return, and to clear the company's name.

Chief Factor James Anderson, who was now running the Mackenzie District out of Fort Simpson, would not grant permission. Campbell then decided to go to a higher authority and thus started yet another incredible journey.

At the end of November, as soon as the ice on the Mackenzie was solid enough, he headed up that river by dog team to Great Slave Lake, Fort Resolution and on to Fort Chipewyan, where he arrived on Christmas day. A week later, after New Year's day of 1853 he continued, still by dog team, via Ile-á-la-Crosse, Carlton House and Fort Pelly, to arrive at Fort Garry (Winnipeg) on February 23rd. From there he went down the Red River and arrived at Crow Wing, Minnesota—the head of the railway—two weeks later. He had walked on snowshoes more than 4,800 kilometres from Fort Simpson.

But when he finally arrived in Montreal and went to see the governor to explain the reason for his trip, he was again turned down. Instead of being sent back to Fort Selkirk as he wished, he was sent on leave. He returned to Scotland, bitterly disappointed and never again returned to the Yukon. When he did return to Canada he served out his years with the HBC as chief trader in the Athabasca District, at Fort Chipewyan and later in the Swan District in Fort Pelly.

The reasons for the refusal to be returned to Fort Selkirk and to continue his explorations there by both James Anderson and the governor himself, is made clear in a report by James Anderson in 1858:

Regarding the enterprise up the West Branch of the Liard and across the mountains . . . I shall merely state that they were utterly hopeless . . . the posts were a source of loss of life, loss of property, loss of reputation, starvation and cannibalism, proving how utterly uninformed were the glowing representations by which the Governor and Council were induced to undertake these enterprises. Dease Lake could only make furs at the expense of

> *New Caledonia, Frances Lake and Pelly Banks made their returns at the expense of Fort Halkett, Norman and Good Hope . . . I therefor trust that the Governor and Council will not again be deluged by any representation by whomsoever made, to extend their posts to the westward beyond Fort Halkett and the Youcon.*

The report, however, did not diminish the fact that Robert Campbell in his travels up and down the Liard had accomplished some incredible feats—feats we modern-day canoeists on the very same river could only marvel at. The same James Anderson writes:

> *In 1842 Frances Lake was established by Campbell. Left Simpson 27th June . . . and reached Frances Lake August 13th. Total voyage, including stoppages, 49 days. Water high. In the fall of 1844 Campbell took 44 days to go to Frances Lake from Simpson. 10 Indians deserted which retarded him but water was very low.*

Thinking about these tough, resourceful men—explorers and their professional paddlers—travelling up and down this raging river in high water and low, in crafts far more fragile than our own canoes, while experiencing the very same canyons and rapids, made all our tasks suddenly not half as tough as we first might have thought they would be. Facing the very same perils of this river in much more solid canoes, with modern supplies and the best available life jackets, in rain gear and warm down bags, gave us a perspective of their life that a mere reading of their journals would never have allowed. Our admiration of these men, their physical and mental toughness, grew daily.

CHAPTER 6
French Canadian Voyageurs

Wednesday, 6th. It was not yet two o'clock in the morning when we made a move against the strong wind, which toward daylight increased to a gale, with heavy rain, and which rendered our landing on one of the points within a few leagues of the river La Loche, a matter of difficulty. Breakfasted early, and waited for a moderation of the weather till about ten when we again took to our canoes . . . Commenced McLeod Portage near five and reached the upper end all safe before eight.
—Archibald McDonald, in *Peace River—
A Canoe Voyage from Hudson's Bay
to Pacific by George Simpson in 1828*

F OR US MODERN-DAY VOYAGEURS the morning after the Cranberry Portage may have been a miserable one, up to the waist in the cold, surging water. But we also knew that we could handle the rapids and the portages and were in good enough shape to enjoy the hard work. Still, when I think of the stories I had read about the voyageurs over the years it was clear that we could never do the things they had done.

The portage certainly had been a struggle—although the heaviest of our packs was no more than forty-five or fifty kilograms. The voyageurs, on the other hand, had always carried at least ninety kilograms (two pieces of ninety pounds each), and some of them much more.

So, who were these men who had accomplished such feats? Who were these men of whom only a very few are known by name? Of whom only a few could even write their name and of whom Hugh McLennan in his book *Rivers of Canada* says, "But the true and proven facts concerning the life of the voyageur are such that I can only say that if I, physically, am a man, he physically, was a superman."

On the trail along Cranberry Portage we had walked in their footsteps and experienced firsthand some of the rugged country they too had come through. What follows here is a tribute to these hardy canoe men without whom the fur trade could not have taken place—without whom the portage trails along the Liard would not have existed for all latecomers to follow.

The story of the voyageur goes back to the beginning of the 17th century. It was then that the first young Frenchmen came to this country that is now called Canada. Men, like Etienne Brulé, who was the first white man to see the better part of Ontario and what is now the state of New York. However, if one were to look for a specific name in connection with the beginning of the profession of the voyageur, it would be Champlain, the founder of Quebec in 1608. With him came the first "young men," as he called them, young men who later became the first voyageurs and coureurs de bois—a unique Canadian word that describes men who travelled the endless woods making contact with the native Indians and establishing routes for the fur trade.

Etienne Brulé too came with Champlain. He was then exchanged for an Indian whom the explorer later took back to France. Brulé was thus actually a hostage but at the same time was exactly where he wanted to be, among the Indians. He was living with them learning their languages and dialects so that later he could act as interpreter. Brulé lived with the Indians for twenty years, became almost Indian himself, wore the clothes of the Hurons, learned their customs, fell in love with their women and, above all, learned to handle a canoe.

In 1615 France began to build her empire in North America. She left the shores of the St. Lawrence, pushed inland, and attempted to take over the fur trade which the Indians had going with a few English, Dutch and Swedes in the state of New York. Brulé returned to Quebec and accompanied Champlain as his official interpreter on his travels to the land of the Hurons. Now his ability in canoe travel paid off, for the entire trip was undertaken in fragile birch-bark canoes. Champlain was one of the first to recognize that if he and his men wanted to travel through this forested country, they would have to forget about horses and European ways of river travel. From Lachine (Montreal) he wrote:

> *The water here is so swift . . . that it is impossible to imagine one's being able to go by boats through these falls. But anyone desiring to pass them should provide himself with the canoe of the savages, which a man can carry easily.*

The canoe route which Champlain and Brulé followed was not an easy one. It led from Montreal upriver along the Ottawa River, from there into the Mattawa, then across Lake Nippissing and down the French River into Georgian Bay. From there their travels took them over a network of lakes and rivers into Lake Ontario and south to where now stands Syracuse, New York.

Brulé, still in a canoe, later explored the Susquehanna River to its mouth, from there returned to the Huron country and saw, as the first white man, all of the Great Lakes. When he was forty-one years old the Huron murdered and ate him. The reason for this rather tragic end never became quite clear, but it is probably safe to assume that he simply stole one maiden too many. Brulé and his colleagues left a number of sons who later began to trade with the white men. And from their ranks came the first voyageurs. Others grew up along the St. Lawrence River—mainly around Trois Rivières—and still others were Orkneymen.

It is always surprising to realize just how much exploring these early Frenchmen did in Canada. Many Americans believe even now that their west was largely unexplored before the coming of the Mountain Men, wandering up along the Missouri. The fact is, that long before that French Canadians had explored large parts of this country. When Francis Parkman in 1846 came over the Oregon Trail, for example, these early explorations by Canadians had already ended. The guides whom Parkman hired in the valley of the Missouri were French Canadians and they were the last of a long chain of canoemen and coureurs de bois, long famous for their canoeing abilities, their stamina and endurance.

David Thompson mentions them in his *Travels in Western North America, 1784-1812*, edited by Victor G. Hopwood: "By the time of the conquest of Canada they had spread themselves far westward, but the Illinois River was their favourite." He then goes on to say that there were about 350 of them at first. A precarious way of life had reduced them to 150 at the cession of the country by Spain. These then crossed the Missouri River and continued to advance to the west, towards the mountains where Thompson first met them. By then, around 1800, there were only some twenty-five left and in 1812 the last two also died a violent death.

As Hugh McLennan wrote, Canada is one of the few countries that was not dependent on the horse for its early development. The country instead was perfect for the birch-bark canoe. With it you could go wherever there was enough water to float it. In addition, the canoes were so light that even the largest one, up to twelve metres long, could be carried by four men.

Once Champlain and his men had started the fur trade, and he and his paddlers began to travel the waterways that led forever to the west and north, there was no stopping them. With a swiftness that today still surprises historians, the canoe, paddled by French Canadian voyageurs reached the country's heartland.

The feats that these professional paddlers accomplished, while moving ever westward, the tough life they led, were nothing less than astonishing. Speed and efficiency were of course essential if the fur trade was to pay for itself. And the schedules these paddlers kept are often hard to grasp today. In brigades—groups of three or four canoes—led by an experienced guide, they left Lachine (today's Montreal) in early May.

Look at a map of Canada and follow the paddlers.

First they paddled, poled and portaged up the Ottawa and then fought their way up the Mattawa. They turned south, crossed Lake Nippissing and reached Georgian Bay via the turbulent French River. Now they headed along the North Channel of Lake Huron, past Manitoulin Island, on the calm waters of the lake, or, more often, fighting strong headwinds. The French called the wind *La Vieille*—the Old Woman. Whenever La Vieille blew too strong, they had to find shelter on shore or risk losing their canoes in the high, steep waves of the lake that acted as if it were an ocean. Good weather or bad, they were expected to complete this portion of the trip in about forty-five days, when going to Fort William (now Thunder Bay). Their canoes were heavily loaded with trade goods and both goods and crafts had to be manhandled across thirty-six portages. Of these at least half a dozen were longer than four kilometres. Express canoes without cargo were sometimes used for important dispatches and they were faster still.

What these canoe travels involved, in hardship endured and in stamina necessary, we will never clearly understand. After all, what does it really mean to somebody who has never paddled a canoe, when I say that we ran some Grade III rapids, that we lined the canoes because the stream was too rocky or too shallow, that we tracked upstream for two weeks, that we portaged seven kilometres through a fresh burn and crossed a four hundred metre high mountain ridge while doing so, that it rained for three solid days, that there was snow on the ground when we loaded the canoes at five a.m., and that we were up to our hips in the cold water because of the rocky shore. It is only by doing these things that you can come to understand them. To get a feel for the hardships these voyageurs encountered, I suggest you read the book by Archibald McDonald, *Peace River—A Canoe Voyage from Hudson's Bay to Pacific*, that was first published in 1872.

True, the journey described in that book was made in two light canoes, especially built and adapted for speediest travel, and they were without trade goods and paddled by nine men each, but what these men did is still incredible.

Loaded with food and equipment for the men and passengers—among them in this case the governor of the Hudson's Bay Company, Sir George Simpson—they covered a distance of 5,217 kilometres (5,089 kilometres in the canoe, the rest in portages) in ninety days. Of these ninety days sixteen were spent at various posts, and a further nine in waiting by the canoes. The actual travelling days were a mere sixty-five. It works out to an average of nearly eighty kilometres per day.

Consider the notes the author wrote on the second day out of York Factory:

> *Sunday 13th of July. Within a few minutes of two the call was given, and precisely at the hour, were under way. With very little exception the men, the whole day and yesterday, on the line. [Pulling the canoes upstream from shore on long lines]. The beach is fine and dry, but water remarkably low. Breakfast at eight, dinner at one, which last stoppage merely occupies from eight to ten minutes, that the men may swallow a mouthful of pemmican, while the servant cuts off a slice of cold something with a glass of wine, to which the governor invites one of his fellow travellers to partake of, as we move along and spend the remainder of the*

day on board his own canoe; the other makes a very good shift to eat and drink something of the kind alone.

From two in the morning to eight at night, day after day.

For a trip from Lachine to Grand Portage at the west end of Lake Superior, the canoes loaded with trade goods were paddled by a team of sixteen to eighteen men. This type of canoe, the Canot de Maitre, was built from birch bark, laid over ribs and planks of cedar. The seams of the bark pieces were waterproofed with a gum mixture. It had to be reapplied daily and when whitewater was run, several times per day. The usual load for this canoe was one hundred and twenty pieces of about forty-five kilograms (approximately ninety pounds) each.

The portages required the most strength and stamina. Here the loads were divided and carried by the crew across watersheds and around rapids too violent or too shallow to be run. Each man, depending on the size of the crew, had to carry two or three hundred kilograms (roughly 600 pounds). Burdened with these loads the voyageurs dogtrotted—doubled over by the weight, with the tump line across their foreheads. Also consider that most canoe voyages took place during the mosquito season.

Almost as tough as the portages was the tracking of the canoes upstream. It could be a nightmare. Here the men hauled the canoes on long lines along the shore often up to their knees in the icy water, up to their ankles in mud and silt, running across lose gravel slopes, and endlessly climbing over deadfalls, trees that had toppled into the river from shore.

Running the many rapids downriver though, was probably the most dangerous work. It was also the most exciting. Understandably not wanting to portage any more than was absolutely necessary the voyageurs often took calculated risks. As a result many drowned. Crosses once marked the places of these accidents. It is said that in some places there were clusters of them, as many as thirty in one spot.

After all this hitting open, smooth water on a lake was a relief. Now the men sang to keep the rhythm and for the sheer joy of it. They also made up time and distance. They often paddled up to 160 kilometres in twenty-four hours with only three hours of rest.

In Grand Portage and later Fort William, the *vrais hommes du nord* (the real northerners) took over. They had come to the meeting place with the furs from the northwest—such as the furs from the Liard River. These men were the elite of the voyageur corps and returned westward in their Canot de Nord. They carried about half the weight in cargo. Much more manoeuvrable they also were easier to carry across the portages. Through a series of lakes and rivers they reached Lake of the Woods and then came, via the feared Winnipeg River into the large inland lakes—Lake Winnipeg and Cedar Lake. Up these lakes they went to the mouth of the Saskatchewan River, then up the Sturgeon-Weir River, across the Frog Portage into the Churchill River, over the famous Methye Portage—nineteen kilometres long—until they reached the Clearwater River. Down it they went to the Athabasca and finally on to Fort Chipewyan located in what is now Wood Buffalo National Park.

Fort Chipewyan was for many years the northern base of the fur trade. From here there were again many possibilities—Peace River, Fraser River and Columbia River. Or, as in the case of the voyageurs who went up the Liard, via the Slave River into Great Slave Lake and down the Mackenzie to Fort Simpson, then up the Liard to Frances Lake and the destination, for a few years at least, was the Yukon.

It was a tough trip and the time table was critical. If the Montreal canoes arrived even a few days too late, it could disrupt the work of an entire summer.

The paddlers knew no luxury. They paddled at a steady pace of about thirty or forty strokes a minute, sometimes increasing that pace to sixty when required. They did this for eighteen hours a day, day in day out, week after week. To sustain this work they consumed between 5,000 and 6,000 calories per day, that is as long as the food was available.

On the eastern half of the route the paddlers were called *mangeurs de lard* (pork eaters) which gives a clue to the type of food that was their staple. With the pork they had corn and peas. In the west the main food was pemmican. Dried pounded buffalo meat mixed with fat and berries was still not much to look forward to after sixteen or eighteen hours of paddling. When there was time along the route which was not often the case, the men fished and hunted and when provisions grew meagre, everything went into the pot—including crow!

I suppose each voyageur had his own reason for picking such a tough trade. All knew that their chosen occupation would eventually cripple them. There were few who reached the age of forty without serious back problems or double ruptures. Rheumatism, a result of wading in icy water and sleeping in damp clothes was always a problem.

So why did they choose such arduous work? One of the reasons was, undoubtedly, the quest for personal freedom. Voyageurs also took pride in their canoeing ability. As historians have pointed out, even though the work was incredibly hard, they were still better off than some of their social class in Europe, or in the lumber camps of northern Ontario and Quebec.

The voyageur was proud of his trade. Even though he did not particularly enjoy a bath en route he always cleaned up before arriving at a post and changed into fresh clothes. Personal freedom demands a high price. For the voyageurs of the fur trade, for the men who accompanied the explorers of the Liard River that price was hard work, a sore back, the risk of drowning and debilitating illness in their old age—for those fortunate enough to survive!

Walking in their footsteps along the Liard River was a humbling experience.

CHAPTER 7

In the Footsteps of the Fur Traders

From Mud River the Liard bends to the North, and still running with great rapidity and breaking into the occasional riffles, reaches, in a couple of miles, the Mountain Portage Rapids, one of the worst rapids met with on the trip. The river here falls over a band of shales irregularly hardened by a system of dikes and worn into a succession of ridges and hollows, and the roughened surface thus produced throws the hurrying waters into an incredible turmoil. We landed at the head of the rapid on the right bank and were forced to spend a day in making a difficult portage of about half a mile with boat and outfit.
—R. G. McConnell, *Geological and Natural History Survey of Canada,* (1887)

DAYLIGHT FOR US HAD COME TOO SOON on that day after Cranberry Portage. We would have loved to turn over once more to stay an extra hour in the comfort of our sleeping bags. But Doug and Wink, always up early, wouldn't give us any peace.

First there was Doug and his daily ritual. He took the small wash basin, tapped it lightly against some rock, whistled and sang as he went down to the water. He filled the basin, placed it on some rocks, stripped to his waist and began to wash himself, snorting and puffing with pleasure. Wink, in the meantime, with the help of the dry kindling that he kept under a corner of his bed, got the fire going. He then banged a few pots together just to make sure we heard him, brought the coffee to a boil and within ten minutes called out, "Mush is on."

The rest of the crew tried to ignore these procedures until the strong, delicious aroma of fresh campfire coffee and an ever-present hunger finally forced us to the fire—with about as much desire to wash en route as the voyageurs had felt.

As far as breakfast was concerned though, we were way ahead of the voyageurs and their pemmican. After a few days on the river we ate amounts of food at breakfast that normally would have been enough for about three full days. First there was Wink's mush, a large bowl of porridge with lots of brown sugar and raisins smothered in milk made from powder. "Porridge that sticks to the ribs," Harvey always grinned. Then there were hotcakes with bacon, maple syrup butter and lots of jam, fruit juice and finally two or three cups of coffee. And when ten o'clock rolled around we were hungry again, we pulled into shore to build a fire to boil a cup of tea, and to snack on some cheese or jerky.

It was almost an hour until we were ready to hit the whitewater ahead which Wink and Harvey had inspected right after breakfast.

"O.K., here is what we'll do," Harvey said, shivering in the cold morning breeze. "There are a few high waves over the next half a mile or so, but with a little luck we shouldn't take on too much water, as long as everybody stays on the right hand side. Drop into the eddy behind that rock, then ferry across the main current to hit the eddy behind the first boulder on the left. If you stay on the right too long you'll hit a number of diagonals coming off another rock with a good hole behind it—so watch out. We will regroup in the second eddy and then take it from there. Doug and I will park in that eddy—just in case."

The canoes pushed off and we followed. Harvey was right; there were indeed a few high waves. While loading the canoe earlier I had ripped the zipper on the spray cover. Now a wave breaking over the bow left me gasping with shock. For a moment

I thought we would tip, but an instinctive quick brace helped and we shot into the eddy.

For a while the river grew calm. Three kilometres farther downstream was another portage around Mountain Portage Rapids. McConnell had spent a day there.

The rapids began in a fairly straight stretch of water. Already from a distance of several hundred metres we could see the foaming white waves. Before the trip I had read in one journal that it was sometimes possible to run the rapids on the left-hand side. Like the voyageurs we too felt that anything was better than a portage and had decided to inspect the possibility. So we drifted down the left-hand bank to land well above the first high waves. Across the Liard was the mouth of the Kechika River, the former Mud River. It came out of the far-off Cassiar Mountains. We walked ahead to inspect the rapids on foot.

The first two or three hundred metres did not look bad at all. Almost runnable. But below that, where the river bent to the right and then disappeared between rocks, the dream of an easy day ended. There was too much force in the way the water lashed the black rocks and then shot straight into the air only to fall back on itself, to tumble over sharp ledges and to disappear again between more sharp-edged rocks. A passage in our loaded canoes would be far too risky, especially since we knew the rapids stretched ahead for more than two kilometres and we could only see the first few hundred metres or so. There was only one thing to do, cross the river and smile at the portage where the fur traders had already struggled.

We tracked the canoes upstream for a few hundred metres and then began the long ferry across the rushing current. Pointing the canoes into the strong flowing river and fixing a reference point on the opposite shore, we pushed off.

At first all went well. We seemed to move pretty well in a straight line and only when we reached the middle of the river did Frank and I begin to drift slowly downstream, while the rest of the canoes pulled away from us. I redoubled my efforts. Still we slipped downriver. A look into the wild, clashing waters two or three hundred metres to my left made me dig in harder. Still we drifted downriver. The far shore seemed to be as distant as ever. My shoulders were burning with pain. I had trouble getting enough air. Anger and fear compressed my chest like a vice and the acidic taste of panic welled in my mouth. What was going on?

A hundred metres upstream and far ahead the other canoes were pulling into shore. Now Harvey looked back and his face registered surprise and worry and then anger. "Hey, Frank, paddle!" he yelled. Frank, apparently experiencing difficulties in keeping the canoe lined up and not quite realizing the strength of the current, was only steering. He was using his paddle as a rudder while I pulled my guts out. Harvey's yell snapped him into action and with a few short strokes we shot across the last of the river.

"Chrrrist! Never again such a crossing! I'm taking the stern from now on," I thought. My chest hurt and my arms were like lead. Frank kept shaking his head as he now looked into the current and chalked it up to another Canadian experience. My anger disappeared as quickly as it had come.

A short search led us to the old portage trail. It was overgrown but had been well used once. The axe marks the men who had come before us had left were still visible on some of the old trees. Four hours later the portage was done. We cooked a noonday soup and rested our sore backs.

Now the valley of the Liard opened up. The forest on both sides of the river stretched back into soft, rounded hills. Three kilometres below Mountain Portage Rapids the Rabbit River joined the Liard. It boiled out of a narrow rocky gorge with its waters white and foaming. Ahead was Whirlpool Canyon.

Below the mouth of the Rabbit River the valley started to close in again. The banks became higher and steeper. I had read McConnell's report again during the short lunch break:

In the next mile the river alternately narrows and expands three times, and falls over short but strong riffles at each constriction, all of which are easily avoided, if necessary by making short portages of a few yards in length. The behavior of the water in the dilated basin between the narrows is somewhat peculiar, as it seems, viewed from the bank, to be running in all directions at once, and to be split into a network of cross currents. At the lower narrows three ugly looking whirlpools are formed by the rapidly contracting stream endeavouring to crowed its way through the narrow channel, while the water sucked down by the whirlpools is thrown up a little farther down in huge boils with the sound resembling of distant thunder.

At a safe distance above the first whitewater we pulled into a quiet bay on the right hand shore and tied the canoes to a couple of large, polished white trees that the last high water had deposited halfway up the gravelly bank. First we followed the shoreline but when a vertical rock face dipped into the river's churning waters, we were forced to climb the bank and then hack our way through thick underbrush along the top. Finally we came to a grassy slope, high above the last tight gate where the Liard widened out to yet another large bay.

The view from here was spectacular. It was also frightening and revealing. Across the river, along the left shore behind a small heavily wooded island that split the swift current into two channels, huge standing waves a metre and a half high were marching by a vertical cliff. It curved toward the centre of the river where the water was then sucked into the whirlpools that McConnell had mentioned in his report. Running that side would be impossible. In the middle of the river directly below the island where the two channels surged together again, almost two metre high standing waves also marched downstream for half a kilometre or more. We knew that although the water would be deep there the waves were simply too high for our loaded canoes. Furthermore they also emptied into the whirlpools below. That left the right hand side and from where we stood, things did not look too good there either.

Another unexpected and long portage seemed to be our only recourse.

While the upper part of the channel seemed to be runnable enough, the lower half was studded with rocks. To make it worse, all the water seemed to be falling to the left, to be sucked, once again, into the hungry jaws of the whirlpools. Only when we finally climbed down to the rocky ledge that jutted far into the stream to cause the narrows, did a passage become clear.

We had to come tight along the right hand shore, backpaddling to slow the canoes down to the speed of the current, thus gaining more time to manoeuvre. We had to shift back and forth and then had to catch a clear chute between two large rocks, one of which was just underwater and not easy to see. But if we came through that all right, we still had to clip the rocky ledge on which we stood as close as possible to catch the eddy behind it, so as not to be washed too far out into the main current that fell towards the whirlpools. The speed of the current would be our greatest enemy. We could not risk to drift too far out and it left little time to line up the canoes. We had to hit that chute just right. But it was worth the risk; facing another long portage seemed far worse than risking a swim.

Nobody said much on the way back to the canoes. Instead we all tried to concentrate and memorize the exact route. We knew only too well that obstacles in the river always looked much different when seen from the deck of the canoe than they did from shore.

A half hour later we fastened down the spray covers. Dale and Adolf would take the lead this time. They would park in the eddy below the outcrop, ready to assist any canoe that got into trouble. One by one the canoes pulled out. The first had already disappeared behind the rocks when Frank and I peeled off. We shot out of the eddy instantly surprised by the speed and strength of the current, and by the force of the wash that pulled us toward the middle of the stream. Any move needed had to be made well in advance in these waters.

Wink and Jeff, in the third canoe, were now between the rocks in a perfect line. But Frank and I were just a little too far to the right—we would end up on the barely submerged rock if we did not correct ourselves instantly. I threw myself far out over the left gunnel and, in a high brace, drew the canoe to the left until the bow passed the rock, only to go into a strong pry so as to straighten out the canoe. "Luck prevails," I thought as the dark shape of the rock slipped by. Then came a loud, sickening crunch. The forward motion of the canoe was too fast and Frank had not been quite quick enough with his pry. The stern hit the rock thirty centimetres from the end of the canoe.

For a moment the canoe lurched and tipped dangerously but we threw a low brace and it steadied. Then we were too busy staying upright and to get into the eddy, to think much about the canoe. But the water pouring in could not be ignored for long. In seconds it sloshed in the bottom up to my ankles.

Luck prevailed again as we made it to the eddy and then ran the canoe up on the sand. Without saying much we took off the cover, unloaded and turned the canoe upside down. Thirty centimetres from the stern the rock had splintered the cedar strips underneath the fibreglass cloth and torn a hole the size of a fist. Its edges were splintered and jagged. But this we could still handle and laughing nervously we relaxed.

Several layers of strong duct tape, friend of all kayakers and canoeists, took care of the hole for the moment. Proper repairs with fibreglass cloth and resin would come later when we had a chance to dry the canoe.

Below Whirlpool Canyon the river ran smooth and fast. On the left Coal River, a clear cold stream that came out of a wide, flat valley, now joined the Liard. Farther back in the valley, fifteen, perhaps twenty kilometres away, was a range of round treeless hills. The nameless peaks belonged to the Terminal Range, the last peaks of the Rocky Mountains.

Below the mouth of the Coal River, we knew, was another obstacle—another canyon, four kilometres long. Judging by all available accounts it was a canyon that could not be run but had to be carried around by a three kilometre long portage. This was Portage Brulé Rapids, the last wild water in the first part of the trip.

With strong, quick strokes we ferried across the mouth of the Coal River where the Liard made a sharp turn to the right. We knew from our maps that the river bent to the left again behind a small island. We wanted to be on the inside of that bend so as not to be carried out into the main current and thus into the rapids below. Rounding the fast-flowing bend, we jumped into the knee-deep water amongst round boulders to walk the canoes to the middle of the bend from where we could see the beginning of the rapids. Most of the river was already falling across the centre of the stream where it crashed against a series of jagged boulders the size of cars. They stood like huge sentinels guarding the entrance to the canyon. From the boulders the water was thrown back in long, large diagonals. It was a spot we needed to avoid.

Already we could hear the thunder of the clashing waters ahead. Two hundred metres downstream stood more dark, dripping boulders crowned with piles of polished driftwood. The water crashed against them with relentless fury. Beside the rocks a seemingly solid wall of whitewater spread to the rocky bank of the right-hand shore and jagged ledges ran across the full width of the channel. Below and to the left of the boulders a rocky ledge ran into the middle of the current from the left shore. It too was throwing the water into a boiling, seething turmoil.

Below the ledge on the left bank a large bay had been formed from where a narrow, flat valley climbed gently toward the wooded hills beyond. We knew by the lay of the land that this was the place

Canoes approaching Portage Brûlé Rapids—Liard River.

where the portage trail started and the bay was an inviting place to land.

We squeezed into the seats to run the section down to the bay. It was not, however, quite as simple as it looked. The whole river fell off toward the middle of the rushing current and we had to work hard to stay close to the left shore. Landing wasn't easy either, the water in the bay was surging up and down more than half a metre among more small boulders. We were finally forced to jump thigh-deep into the water to hold the canoes and to prevent them from smashing into the rocks.

In a few minutes though, supplies and gear lay piled on the dry shore, the canoes turned upside down on two large spruce trees which the high water had deposited on the beach. Here they would dry in the sun for an hour or two to be ready for later repairs.

Camp that night was ideal. The stony beach ran back from the river for about ten metres where it met a strip of white sand where willows struggled to grow in shallow soil. Behind the sand a clay bank climbed to a height of two metres, overhanging in places, and washed out by the flood waters and the snow melt. On top of the bank a strip of parkland with grass, birch and poplar trees ran along the river, before it joined the heavy bush that covered the round hills farther back. Here we put up the tent, built the fire and put the kettle on the boil.

The day's work though, was not yet finished. After a quick supper we shouldered a light pack, grabbed axes and began to search for the portage trail. Down by the water's edge a large old spruce bore the scar of a wide blaze, and farther inland, where the dense bush began, we found another old blaze. The beginning of the old, well-worn path looked good at first. It followed the narrow valley, turned toward the river, cut across a rocky ridge and then began climbing steadily and easily inland.

We followed the clearly visible trail for a few minutes but then soon something did not feel quite right. The trail led too far inland and there were no more old blazes on any of the trees. Instead we came to a pile of fresh moose droppings and we knew we were on the wrong path. What we were following was simply a well-travelled game trail that ran down to the river from the hills beyond.

Dusk was turning to darkness, too late to look farther. Hanging the packs on low branches of trees, to be picked up the next day, we started back to camp. Halfway there we spotted the place where the real portage turned off the game trail. It followed a semi-open sidehill and then climbed out of sight. Harvey and Adolf turned into it to follow it farther and to get an idea of the conditions we would meet the next day. "Just for a few minutes," they said.

Back in camp there were still things to be done. Bannock had to be baked, the canoes needed repairs, firewood had to be cut and my journal needed updating.

Slowly the northern darkness settled in. The dark, dripping boulders in the centre of the stream began to blend into the dark outline of the spruce forest on the far shoreline. Soon only the distant roar of the water indicated that there were rapids downstream. Harvey and Adolf had not yet returned. We began to worry. The heavy bush of this region was not the best place to be after dark—especially not when one thought of the grizzlies which McConnell had mentioned in his report.

It was now eleven at night. Frank, Jeff and Dale had long ago crawled into their sleeping bags. But the rest of us could not sleep. We sat by the fire with steaming cups of tea and an open can of honey. Finally we heard voices and then the breaking of bush—but it came from the wrong direction. Grinning somewhat sheepishly, Harvey and Adolf stumbled into the fire light. They had, they now told us while sipping a hot cup of tea, somehow lost the trail in the dark, had then wandered too far inland, bypassed our camp and had only regained their bearings when they found the bank of the Coal River, far upstream. They had then followed it to its mouth and came back along the shores of the Liard.

Despite the late night we were up early the next morning. The day would be a tough one. What we really needed was a day of rest and not another long portage; still things went well for the first few hours. We relayed packs and canoes in easy stages down the freshly blazed trail that Harvey and Wink had scouted, following the lay of the land. Deadfall, as usual, was the worst obstacle and thick stands of young poplars sometimes seemed to entangle the canoes. Soon, all of us were resigned to the grunt of heavy packing and only Doug was still whistling. Things became monotonous.

Shortly before noon I met up with Harvey sitting on one of the packs. He looked terrible. His face and neck was red and swollen, his eyes almost shut, his whole face a mask of pain. "What the hell happened to you?" I asked.

With the same sheepish grin of the night before he tried to laugh, found it too painful and said, "There won't be any more blazes for the next couple of hundred metres or so—at least none that I made. I stumbled over a deadfall a ways back, and fell headlong into a hornet's nest. I took off but they still got me. Anyway, you best turn off the trail when you reach the last blaze just up ahead."

Later Wink also had a story to tell. He had moved too far

inland and had entered another draw. Only when he realized that he couldn't hear the river any more did he know that he was turned around. Only his compass helped him to find his way back to the river. "Believe me," he laughed, "I had to use all my willpower to follow the heading of the compass—I was convinced it was pointing the wrong way."

We cooked a pre-packaged chicken noodle soup and then turned off the trail to bypass Harvey's hornets. The packs soon grew heavier, the canoes tougher to handle. The trail seemed endless.

Halfway along the portage on a semi-open terrace, not far from the riverbank, we came upon the ruins of three old log cabins. Their pole roofs had long fallen down and trees grew inside the walls. Judging by the size of these trees, the cabins were at least forty or fifty years old. In one of them stood a perfect fireplace built of flat, brown rocks. Whoever had built these cabins had done so with pride and skill. Maybe some of the men who came up along this river during the gold rush to the Yukon had wintered here. We did not know it then, as we came along this trail the first time, but the builders of these cabins had had an excellent reason for wintering on this particular spot.

Only two years later did we discover their strategy. We were clearing a new portage trail, this time running it along the river's bank. Adolf who with two of his sons was doing most of the work with chainsaw and grubhoe, stumbled onto a number of hotsprings while following a moose trail to the water. For more than three hundred metres along the bank, hot mineral springs oozed out of the rocks in many different places. Some, flowing out of the gravel seam halfway up the bank, were so rich in minerals that their deposits over the years had formed high mounds of tufa—perfect sunken bathtubs. Other springs flowed out of the bank at the river's edge. There it was possible to have one buttock in the cold of the river, the other in the warmth of the spring, as a few of us soon tried to demonstrate. It did not take long that day for the clothes

Jim Fornelli at remains of cabin near Hotsprings at Portage Brûlé Rapids.

to come off and for six men to lie back lazily in the warm water, while idly watching the furious waters of the Liard shooting out of a tight gap in the gorge and then disappear over sharp rocky ledges.

But that still laid two years in the future. Right now we were merely soaked in sweat and did not even dare to dream of a hot bath.

From the front of the cabin we looked down into the rapids of Portage Brulé. Soon we were no longer wondering why we were packing gear and supplies over solid ground. Large, jagged, shaly boulders often ten, fifteen metres long, cut the river into numerous channels. Each boulder was crowned with a tight wedge of solid driftwood, deposited there by high water. Between the boulders bands of rocks, sometimes just under the surface, angled across the entire width of the river creating a series of steps. At the bottom of these, large vertical eddies, or holes, churned the water into an airy froth. These were the dreaded "keepers" large enough to swallow whole canoes and anybody unlucky enough to be in them. Where the channels met again, below the boulders, huge standing waves marched downstream only to be thrown against still more rocks. Nowhere, as far as we could see, was there a clear chute or channel through which a canoe could have been run with any degree of safety.

This knowledge though, did not make the portaging any easier. We laboured on. On a few of the faces exhaustion began to show. Nobody cracked any jokes. Only the next step was important, and then the next, and the next....

After more than eight hours it was done. Packs and bundles were piled on a wide gravel bar where the Liard waters again ran smooth and fast. At the end of the portage we used long ropes to let the canoes down over a steep grassy slope. Here, when McConnell had come this way, rollers and a windlass had still been rigged to help the voyageurs and later the miners, drag up their heavy boats. There was nothing left when we came through. Here too scattered on this slope, Robert Campbell had found the packs of the outfit which, the summer before, had been on the way upriver destined for the new post at Dease Lake. (Campbell had found the goods where the men had dropped them, after they had panicked by silly rumours that a band of "Russian" Indians was approaching.)

It started to rain lightly when we put out the fire we had used to boil a pot of tea and then reloaded the canoes. Ahead lay two days of easy paddling without rapids or portages, filled instead with encounters with moose and bears.

Right now though, it was time to find a good campsite. Two kilometres farther downstream we found it. A group of tall spruce trees high on the right-hand bank, their branches spread wide, offered protection for our tents, and a sandy beach beside a small creek provided a good landing spot for the canoes. We pulled in and unloaded.

It took only a few minutes to get a huge long fire going, despite the drizzle. Stuffed with a large meal we stretched out around its warming blaze. Soon the hardships of the day grew unimportant. Talk turned to the Klondikers who had come this way in 1897 and 1898—in the midst of winter, as their diaries told the story, in temperatures of minus forty and colder. They struggled upriver on their way to the Yukon. Sick with scurvy, snowblind, and with frozen limbs, they fought to survive.

"Hey, would any of you guys want to do that," Adolf asked, lying on his side by the fire, his head propped in his hand.

"Nah," Doug answered, "but give me a side of bacon, some salt and some sugar, a few bags of tea and a canoe and I'll be content for quite a spell—after all who needs the gold, eh?"

CHAPTER 8
The Gold Seekers

February 1898, 10th, Thursday: George better today. Caught in storm this morning, frostbitten again. Took six hours to get to the freight pile instead of 2 and a half—trail snowed up. Did not regain camp till dusk, very heavy work hauling toboggan, snow often 4 feet deep....

March 1898, 10th, Thursday: ... We do not get encouraging news of the Yukon country: 2,300 men went in last fall and 700 turned back disgusted. We have made a great mistake coming up the Liard, it is the hardest way to get to the Yukon. We might have started this March in Telegraph Creek, and then be further ahead than we are now, after a hard winter's work and most of our grub gone.

—Alfred E. Lee, *Diary of an Expedition from Edmonton 1897-1898*, annotated by T.L. Brock

THE MEN WHO CHOSE THE LIARD route to get to the Yukon were, of course only a small contingent of the men and women who eventually reached the creeks of the Klondike. This route, however, was without a doubt one of the most difficult.

This "All-Canadian-Route" as it was called, was first promoted by a certain Arthur Hemming, an artist and woodsman of the time, and by many others—among them the editor of the *Edmonton*

Bulletin. "This route," he wrote, "4,000 kilometres long, follows the trail of the fur traders and is by far the cheapest. All that is needed is a good physical condition, some experience in boating, a tent and about 50 dollars. It can be done in about two months."

The good man blessed with his vivid imagination, forgot to mention a few rather important details. One was the fact that, though the voyageurs of the HBC did travel the route not in two but in three months, they used light birch-bark canoes, not heavy wooden boats. Moreover, these men were the best of the entire voyageur corps, well-seasoned in travel under rough conditions in the bush. They were not greenhorns like most of the Klondikers. They also had the HBC backing them. The Klondikers, on the other hand, were entirely on their own and were moving up to a ton of gear and provisions per man.

Then there was the rough water of the Liard. It too was never mentioned and certainly not widely known. Already in the Athabasca River the rapids seemed endless: Grand Rapids, Burnt Rapids, Drowned Rapids, Middle Rapids, Long Rapids, Boiler Rapids, Cascade Rapids. The Slave River too had its rapids and in one place the men were forced to transport their goods overland for more than twenty-five kilometres because of rough water. Once they reached the Mackenzie River and its tributaries, the promised land was still a long way away—beyond snow-covered mountains and beyond solid, totally unknown wilderness.

All the men who picked this route were under way for almost a year, many longer. All fought violent waters, scurvy, freezing cold and the boredom of relaying load after load, day after day, week after week. A few reached their destination; many died on the way.

Fort Simpson on the Mackenzie, 1,800 kilometres from Edmonton was situated about halfway along the route. Here the men had a choice. They could, if they wished, follow the Mackenzie downstream to Fort McPherson, then go up some tributary such as the Rat River, or, they could follow the Liard and get to the

Yukon by what they thought to be a shortcut. J.G. MacGregor, in his book *The Klondike Rush through Edmonton* describes the route:

> *Looked at on any of the maps of the day, the Liard River must have seemed a most sensible way to get from the Mackenzie River to the Yukon. In a straight line Fort Simpson lay some 500 kilometres directly east of Pelly Banks, the point at which boats put into the Pelly would float down the current into the Yukon River. The Liard, however, did not follow a straight line, but swept along in a great wiggly semicircle south, then east, then back to the north again, so that its headwaters in the mountains of the Yukon were 1,100 kilometres from Fort Simpson. And all of its 1,100 kilometres were not only upstream but for the most of them its waters came roaring and whirling and gushing down a series of rapids, cascades and canyons. Deep and broad and almost sluggish as it approached Fort Simpson, however, it presented a beguiling appearance. Moreover, the Klondikers knew that some 350 kilometres upstream at the mouth of the Petitot River stood the Hudson's Bay Company's Fort Liard and other posts were said to be strung out beyond that. Unfortunately, the Klondikers never heard what the Hudson's Bay Company's staff knew about the river, or, if they did, ignored it.*

One of the many canyons mentioned by MacGregor, the Grand Canyon of the Liard, was said to be such an obstacle that even the Indians did not travel it. Below the canyon the Indians looked to the east and traded in Fort Liard, while above it they looked to the west and traded at Lower Post or Dease Lake.

In September of 1897, twenty-five men, belonging to several parties and loosely connected groups, began to track their boats upriver from Fort Simpson. Even for these men the river held many surprises, despite the fact that all twenty-five had plenty of experience in living in the bush—in great contrast to those who followed them—and had sufficient supplies to last them the length of the trip. The trip soon became a test of endurance, a test of the ability to survive. Gone were the lazy days of drifting on the broad Mackenzie, instead there were now days, weeks, and months of gruelling work—work with paddles and oars and tracking lines, with constantly wet feet from wading in the cold water as the men dragged their boats upstream, often struggling for each single metre.

One of the first lured by the promise of easy gold and perhaps the only one who had a real knowledge of what lay ahead, was Charles Camsell, (who, twenty years later was to become Canadian Deputy Minister of Mines.) Twenty years old, born in Fort Liard he was a true son of the north. Together with his brother Fred, who kept a diary of the trip, and four other men—D.W. Wright and A. M. Pelly, both experienced prospectors from the Okanagan, and Dan Carey and his son Willy, also experienced in living off the land—left Fort Simpson in September of 1897. With them they had five or six Indians who helped them track their boat. Five weeks later, by the end of October, they had reached a spot some twenty kilometres below the mouth of the Toad River, still well below the Grand Canyon of the Liard. Compare this with some of Robert Campbell's travels! Here they stopped because the ice on the river was getting too heavy. The Indian helpers were paid off and returned to Fort Simpson in one of the boats. Camsell and his friends then built a log hut and waited three weeks for the ice to be firm enough to walk on.

On the 20th of November, in temperatures that had plummeted to minus thirty degrees Fahrenheit, they set out again. During the three-week layup they had built sleds to which they later hitched the dogs they had brought along. Charles Camsell in his book *Son of the North* later wrote:

> Besides our camp outfit we had about 5,000 pounds of provisions, enough to last us all winter but too much to carry in one load. Our plan, therefore, was to load our toboggans to capacity and

carry this load upriver for ten miles or so, cache it there and then return to camp the same day, making our trip about twenty miles. At first it took over a week to move the whole outfit ten miles. On the last day we broke camp and moved twenty miles for our next camp and from that point as base we brought our supplies up and then carried them forward another ten miles, at least that was what we tried to do, but it was not always possible. Bad weather, deep snow or rough going frequently slowed us down so that we only made a fraction of the distance. Fortunately sickness never bothered us. Accidents were of minor nature though I was layed up for a week when I cut a gash in my knee with the axe. It was slow going at first, but as we consumed our provisions the intervals between moving camps shortened until toward spring we were able to carry everything in one trip. By the first of May when we reached Frances Lake, we must have tramped altogether some 3,000 miles on snowshoes.

How easy it sounds when six months of winter travels are summed up in such few words. It is only when you read Camsell's book between the lines, that you will realize that it was not easy at all, even though he and his friends knew how to take care of themselves and were also well equipped.

Later, the Camsell brothers decided not to continue all the way to the Yukon. They assumed, quite correctly, that all the good claims had already been staked. Instead they spent the next two years trapping and exploring the country around Frances Lake and Dease Lake. Then in 1899 after almost starving a few times, they returned to Fort Simpson in a canoe trip which he claims as one of the best in his life.

Others were not so lucky. Of the twenty-five men who were the first to come up the Liard, only ten finally made it to the Yukon. Four others stayed a while in and around the headwaters of the river, eight returned to Fort Simpson and three died.

One of the men who eventually did reach Dawson City in the

middle of June, 1898, was Alfred E. Lee. He left many of his impressions of his remarkable journey in the form of a diary.

Alfred E. Lee belonged to a group of six men—the others were Bill Howey, T.A. Stephen, Alec Gibney, Harry Woodward and G. Purgess. He began his trip in Edmonton, on August 9th, 1897 and reached Dawson City on the 20th of June the following year. During this time he walked a total of over 7,000 kilometres, including the endless relays of transporting their provisions in the same manner as Camsell and his friends. The diary of this remarkable man, Alfred E. Lee, was published privately in August of 1976. Only 165 copies were printed with Mr. Thomas L. Brock as the annotator of the book.

Alfred E. Lee was also one of the few who stayed in Dawson long after the hectic time of the Gold Rush had come to an end. Only these diary entries can shed some light into the conditions and hardships these men endured during their long, arduous struggle up the Liard in the middle of the coldest time of the year. In Fort Simpson, where Lee and his partners left on the 19th of September 1897, many of the different parties that had come down the Mackenzie River had regrouped and some were now pulling together:

> *September, 27, Monday: Cold weather and more sleet and snow. We get no spare time at all, eat our breakfast and start an hour after turning out in the morning; dinner an hour at noon which time includes cooking; quit towing at 6 or 6:30 p.m., then have supper and bake bread for the next day and wash up it is generally 10 or 10:30 by that time, then turn in. It is pretty hard work, a fellow doesn't get time to wash or anything of that sort, but time is a serious matter with us now.*
>
> *28th, Tuesday: Slow progress owing to sand bars and fallen brush on shore; pretty cold wading in the water first thing in the morning. Camsell and Wright and Pelly outfit in two boats caught us up at noon.*

October, 8th, Friday: We find we can only make about six miles a day. The river is getting very bad, full of islands sand bars and shoals, the water being generally very swift in the shoals. Our boat is far too unwieldly and heavily loaded to enable us to make good progress. When we cross the river, the current often carries us back half a mile. In one place, we only made 3/4 mile in an hour and a half, everybody pulling his level best and catching on to the twigs and roots that he might get a better hold.

18th, Monday: Snowed fast last night. I was rash enough to sleep without any canvasses over my bedding. The consequence was in the morning I had an extra blanket of snow which kept me beautifully warm, only the snow is so soft that it wet all my blanket through. We have not started out today yet (noon) as the snow still continues.

24th, Sunday: Decided to camp here, as there is too much ice in the river to allow us to go any further. In the afternoon Harry, Alec and I walked about 4 miles up the river finding a man named Shaw and his partner camped there. They informed us that Wright and Pelly are camped just this side of Toad River.

November, 17th, Wednesday: 30 below zero last night and 25 below at noon. Fetched 200 lbs more flour from Proposition Island, but it was very hard pulling, the trail was not good and the snow very deep.

21th, Sunday: Shaw brings word that Martin will probably loose two of his toes as he got both of his feet frozen rather badly. Graham also froze his toes. Atkinson and DeWolfe intend going through to the Yukon with dog trains.

December, 2nd, Thursday: Bill Howey and Stephen started for Toad River this morning at 9 o'clock to get intelligence about the river ahead. Alec, Harry and I propose setting out for Devil's Portage overland to prospect a toboggan route, as the distance overland is 60 miles and by river 85 miles from here. We set out from camp about 1:30 p.m. and camped in the cooley beyond the Gap, 10 miles above here, taking our blankets, 300 hardtacks, 20 lbs bacon and 2 lbs tea, beside a little bread we had on hand.

Along the route overland, from which they returned on the 16th of December, they met an Indian family. From them they learned that they had reached a point close to Fort Halkett and were in fact two days past the Devil's Portage. However, Lee found the trail impracticable for hauling freight and decided to stay with the river. Two days later, Lee and his party moved into the shack which Wright and Pelly and Camsell had built and then used it as a dropoff point for their provisions which they moved up during the rest of the month. On January 3rd, 1898, they finally moved on to a new camp and from then on lived in the tent, forever relaying their freight. By this time most of the original parties had broken up, mostly because of friction among the men living so close together for such a long time and new partnerships had formed. Moving up and down the river became an endless routine. They had supply caches in many different places and passed each other constantly. All of January was spent this way. On the 12th of that month, they came to an Indian camp about ten kilometres above Hell's Gate, which is the entrance to the Grand Canyon of the Liard going upstream. Lee was now past the Wright and Camsell camp. But things were still tough as his diary shows:

January, 30th, Sunday: Alec and I fetched the last load from Dan Carey's camp. Dan had decided to stay there till spring and go around the Mackenzie Route to the Yukon. I gave him permission to use my canoe. [Lee had cached his canoe at the abandoned Toad River post]. We gave him two spades, 4 picks

50 lbs bacon, a lot of soap and other things. He gave me his toboggan. There are some tremendous cracks in the ice through the canyon, covered with snow so as to be invisible, and if you happen to step on one down you go. I fell through about 50 times these last few days, sometimes up to the armpits—it is not pleasant. The country that we are now passing through has some terrible names; we first passed through Hell's Gate and are now camped at the Isle de Grave's. Today the other boys took up loads upstream and passed the Rapids of the Drowned while Alec and I fetched the stuff that we had down at Dan's camp. This trip was made in 22 trips (11 miles).

February 10th, Thursday: Caught in a storm this morning, frostbitten again. Took 6 hours to get to the first pile instead of 2 and a half—trail snowed up. Did not regain camp till dusk, very heavy work hauling toboggan, snow often 4 feet deep. All hands about petered out tonight.

12th, Saturday: Took the first load through Devil's Canyon. On the return trip about half an acre of ice we had passed over had been carried away by the water close to the rapids. Pelly and Wright pitched their tent close to us tonight. Very rough and rapid water at the top of the canyon.

Camsell later wrote that the passage through the Devil's Gorge was the trickiest part of the whole winter's travel, but it was still preferable to the hard climb over the portage.

26th, Saturday: All three of us went out moose hunting with Wright party, but did not strike any. Saw the hot springs and lake of sulphur water—it is on the north side of the river about 1 mile back and 3 miles above Trout River.

March, 8th, Tuesday: Hauled 2 miles beyond Coal River; stormy day. Met an Indian with dog train to help Wright and Pelly haul freight.

Lee had now gone past Fort Halkett at the mouth of Smith River and also past the spot where we had found the old cabins along Portage Brulé. On March 13th Lee had had a falling out with the rest of the party, especially with Alec Gibney and decided to travel on with Pelly and Monte Velge. Lee then sold much of his supplies so that he could travel with one load of about 250 lbs.

March, 20th, Sunday: Went about 20 miles, Pelly fell through the ice, and I went in up to my knees. Could not find a camping ground till dark; cold night. Went back to help Pelly with his load and meantime froze the soles of my feet slightly.

21st, Monday: Feet very sore. Reached Silvester's Post (Lower Post) about 3 p.m. and find we cannot obtain supplies as they are quite out, so we have to pull for McDame Creek. Rest tomorrow to give my feet a chance and pull out the next day. Hope to meet Simpson coming up with supplies, no rest for the wicked.

Lee now made a trip of some 180 miles (290 kilometres) to get fresh supplies to last for the rest of the journey. He and his partner started out for McDame Creek on the Dease River on the 24th of March and returned to the Liard on April 1st.

April, 7th, Thursday: Struck Frances River about 4 p.m. Pelly and Indian camped 2 miles back. Water and slush everywhere, always ankle deep and sometimes up to our knees. All this for some gold which we hope and believe exists some 600 miles away. We were glad to bid adieu to old Liard River, and I for one hope never to spend another winter hauling a toboggan up it.

Lee then trudged another month or more to reach the Pelly River via the Finlayson River and a portage across the mountains. On the shores of the Pelly, he and his friends built a boat and then floated down the Pelly and the Yukon to finally reach Dawson City on the 20th of June, 1898. Not all of his friends made it there. In a boat wreck in Hoole Canyon, two of his friends drowned. Just exactly who they were is not quite clear. Lee's second diary, written in a new book, and dated from May 10th to his arrival in Dawson has, unfortunately, been lost.

MacGregor, however, in his *The Klondike Rush Through Edmonton* fills in some of the missing pieces:

> *Eventually, Lee, Woodward, Gibney, Pelly, Velge, Graham, Schreeves and Balaam reached Frances Lake area as two or three groups, while Shand went along—nine men now. By this time some prospectors from the coast also reached Frances Lake and there are new names . . . The nine Edmonton-based Klondikers went whirling down toward the Yukon River, 280 miles away. Hoole Canyon, some forty miles down the Pelly looked dangerous and half of the group portaged around it. Woodward and Schreeves and a man named Kennedy, who had come from the coast and had been weakened by scurvy, ran it in one boat, while Gibney and Balaam started out in another. Woodward's boat came through safely, but Gibney and Balaam were swamped and both drowned. Seeing the accident affected Kennedy so strongly that he dropped dead.*

Lee and the Camsell brothers, who came up the Liard in 1897, with about eighty other men that year, however, were only the vanguard of those who chose the same route a year later. The Liard route then became a route of sorrow, of suffering and of death.

A few more quotations from other diaries—from MacGregor's book—reveal their struggles. Wrote a certain A.D. Stewart:

> *Edward Grundy, from the state of Maine crept into our tent this morning begging to assist him. He had ventured up the river getting almost as far as Fort Liard when his boat swamped and he barely managed to escape with his life. The unfortunate man had on a hat, and undershirt, a pair of blue overalls, and a pair of moccasins, and outside of these with the exception of a little coffee, a pail of lard and a revolver which he had managed to save, the poor fellow had not a rag left to cover him, not a cent of money to take him home, nor any means of going onward even if he inclined to pursue his journey. The fact is we have heard of so many casualties, so many broken boats, so many damaged provisions, so many discouraged people that we pay little attention to anything we see or hear, but our precautions grew greater day by day, and we move more and more cautiously as our journey proceeds . . .*

Another man named Oliver, from Edmonton, had set out in the late summer of 1898. He reached Trout River, where the Alaska Highway now swoops down to the Liard for the first time, on April 5th, 1899. By the end of May he and his party were at Portage Brulé. There they had the following encounter which was later reported in the *Edmonton Bulletin:*

> *When on Brulé Portage a man named Hutton came to them starving. He and his partner, named Knute Nelson, a Dane from Chicago, had taken the cross country trail from Snyetown to Mud River Post. (This was an old Indian Trail from La Biche River, above Fort Liard, to the Coal River and then on to Mud River Post, opposite the mouth of the Kechika River). About forty miles from the post they had left their guns. Mr. Edwards of the HB post there had given them food to go back for their rifles. They had gone back but run out of food. When near the crossing of the Coal River a few miles from the post on their return, Nelson*

gave out and Hutton went on alone. In trying to raft across Coal Creek he was carried down the Liard and only succeeded in landing just above the Brulé Rapids. Two of the Oliver party went with him to the relief of Nelson and found him dead, but still warm. He was buried there.

When Oliver finally reached the mouth of Dease River and heard that nobody was having any luck along Frances Lake and even farther north and west, he decided that he had had enough and went home.

Sounds so easy. But to get home he tracked the boat up the Dease River to the south end of Dease Lake. From there he hiked some 120 kilometres to Telegraph Creek, canoed down the Stikine River to Glenora and from there took a passage to Wrangell and then travelled back to Edmonton. Oliver was more than sixty years old and had endured hardship that many, forty years younger, could not withstand.

In the meantime, others were still coming up the Liard. Many of these new arrivals had come down the Fort Nelson River, to link up with the Liard from there. They kept coming until there were as many as four hundred men scattered along the river, between Toad River and Lower Post. Many of these men were in real trouble by the spring of 1899. A petition which was sent to James Porter (the gold commissioner of the Cassiar mining district, who had his headquarters at Telegraph Creek) shows the extent of the trouble that these men had found. The petition, signed by forty-two men, and dated Mountain Rapids, B.C. April 3rd, '99, reached Porter on May 7th and said:

We the undersigned beg to inform you that there are men on this river, three at this spot and several near here, who are disabled entirely destitute and who unless government aid is sent immediately, must die. We have appealed to the HB Cos agent, but, he refused to aid them in any way. We are ourselves unable to provide them with the provisions or take them to civilisation having all we can to take care of ourselves. We would urgently beg you to send at once such means of transportation as seems best and provide these men with the necessaries of life until such time as they can reach civilisation. The men for whom we ask assistance are: Fred K. White, has scurvy, unable to walk—has no provisions. Alex McCulloch—feet badly frozen will not be able to walk for several months at least, has not more then 15 days grub. Nels Johnson has scurvy—been unable to walk for several weeks—has perhaps 30 days grub. Wm Barker has scurvy, has been pulled on a toboggan over a hundred miles and his comrades, having no grub, were unable to stay with him, so they left him here. He is very weak and will be unable to move for several months. He has not more than 30 days grub. W.H. Harris, feet badly frozen cannot walk for months to come, has perhaps 60 days grub but no means of travelling before it will be gone. In view of the great necessity now existing we beg you to take immediate action.

Fred Camsell, who had come up the Liard almost two years before, took charge of the relief expedition on May 8th. After about two months, when his scow finally returned to the upper end of Dease Lake, they had between fifty and sixty men with them. Some were so sick with scurvy, that they had to be carried on and off the boat at each camp. With the proper diet however, all of them eventually recovered and returned home.

MacGregor estimates that about twenty-six men died that winter in the Liard-Cassiar region, seventeen from scurvy and the others drowned.

Still, others had come through. They had proven themselves. They had pitted their strength and wits against the Liard, against all the mighty river could hand out, and they had won. Long before they reached the Yukon, the gold that they had been after when they started out, had become unimportant. Instead they fought to survive, came to know themselves and became part of a legend.

We who were on the Liard almost eighty years later could not really pity the men who had perhaps built the cabins halfway along Portage Brulé, and who later had perhaps died on the trail. They had done what they had wanted to do; they had followed a dream. Just as we were doing in canoeing this same Liard. For he who abandons his dreams will not suffer only their loss, but the loss of his spirit as well.

CHAPTER 9
Across the Devil's Portage

We had been on the Lookout for the Devil's Portage and Rapids ever since leaving Rivière des Vents, and as the threatening appearance of the river and the valley indicated that we were approaching them at last, we dropped down cautiously along the right-hand bank, watching carefully all the time for signs of the old portage. We failed to discover any, but landed at what seemed the last break in the almost vertical cliffs, with which the river was now bordered, and just at the head of a long, easy riffle. Looking down the river ominous streaks of white could be seen in a couple of places, stretching from bank to bank, while the familiar roar of clashing waters was clearly audible. On landing we found, after a short search, traces of the old portage track, and the next few days were spent in carrying our outfit across.
—R.G. McConnell, *Geological and Natural History Survey of Canada, (1887)*

WE ALL AGREED. The diaries of the Klondikers and the voyageurs made fascinating reading by the comfort of a late-summer campfire. We also agreed that the small discomforts we had experienced that day on the trail around Portage Brulé paled in comparison with the struggles of those determined men. Ours had been a tough day only by our own standards, as would the struggles be that lay still ahead. It gave us comfort to know what limits men could reach when the chips were down.

We slept well that night. Toward noon the next day when we arrived at the mouth of Smith River, light rain was falling like a silver mist from the dim grey sky. An hour before we had chased after a moose we surprised swimming across the Liard to get close-up pictures. And later we had come onto a black bear swimming for the far shore, where it disappeared into a thick stand of willows.

We crossed the mouth of the clear stream that was Smith River and landed on a wide, sandy beach covered with fresh moose droppings and crisscrossed with tracks.

This was the spot where Fort Halkett once stood, perhaps the most important trading post on the upper Liard in its time. Robert Campbell had been in charge here when he went up to the Dease River to establish a new post there. Altogether Fort Halkett had lasted some forty years, from 1822 to 1862. The storms and floods of more than a hundred years after that had effectively obliterated even the last remains of the buildings. The land claimed back its own. Where once were small patches of wheat and barley and vegetables, there were again only willows. The men who had lived here, more often than not at the verge of starvation, in the loneliness of the land in the winter, were all but forgotten. Only Robert Campbell's journals and the records of the Hudson's Bay Company still tell their story.

We followed the river up its course for a few hundred metres but managed only to spook two moose that crashed out onto the beach and then swam the river.

After leaving Smith River and the ghosts of Fort Halkett, a strong, cool wind rushed down from the slate grey clouds, blew straight up the Liard and cut through our clothes like a scythe. This was the same wind that decades ago had tormented the voyageurs when they named a tributary in this region Rivière des

Vents—Windy River. It spilled its waters into the Liard on our right, clear and dark green.

Battling the strong headwinds slowed us down and it was mid-afternoon before we came around the last long bend to see the suspension bridge of Liard Crossing that carries the Alaska Highway across the river. We pulled in and put up camp under the bridge. Soon a large driftwood fire was drying our clothes and gear.

It was still raining, when, an hour later we set out for the Hotsprings a short distance away. The springs were first mentioned by Campbell in 1837. They are now a well-frequented tourist attraction in a provincial park known as the Liard Hotsprings Provincial Park. But more important to us, they were exactly what we needed after a hard week in the bush.

The small pools were overcrowded. Somewhat naive I had pictured them as hotpools tucked away in the wilderness and had not even bothered to bring along swim trunks. But the long line-up of cars and trucks along the edge of the highway soon destroyed any notion of skinny dipping. Luckily Harvey and Wink had brought along their swim trunks which we now shared by changing in the small smelly change room.

It seemed that all the tourists of the Cassiar country had gathered here to take a bath and I had some difficulties in adjusting to this sudden onslaught of human flesh. Especially since it did not represent itself in its best form. Women in all stages of age and development—they too had obviously forgotten their swim gear—stood in panties and bras in the hip-deep water, soaping themselves and washing their hair, until the lower portion of the pool was a milky white slime. Men, with the excess fat of civilisation, sat on a log anchored in the middle of the pool, ogling the women while slowly turning red as boiled lobsters. But the water was warm and soothing, and where it flowed from the rocks at the head of the pool it was hot and clear. It was a good feeling just to lie back, to soak up the warmth and to half-lustily watch the women

Consulting maps on the upcoming course of the river.

who tried their best to appear nonchalant when the water rendered their nylon bras and panties transparent. Today's condition of this provincial park is much improved. The changing rooms have been cleaned up and rebuilt, the use of soap and shampoo in the pool is forbidden and the slimy bottom of the pool has been covered with clean gravel. The park now also has a tenting area with firepits and stacks of firewood.

We finally returned to our camp under the bridge, more tired than before as the muscles were relaxed from the hot mineral waters, and decided to stay put for the night. For once, we thought we could get away without having to put up the tent. Later in the evening though, and then throughout the night, when a storm came up that transformed the shelter under the bridge into a hellhole of whistling winds, we regretted that decision. The smoke from our cooking fire whipped through the girders of the bridge and out along the river as if driven by screaming witches.

After a late supper Adolf took the detailed maps of the Grand Canyon of the Liard out of their watertight container and laid them out on the ground held down by small rocks. Judging by these maps that had been drawn in the early sixties, after surveyors had pinpointed potential dam sites, it was clear that the most difficult stretch of the river was still ahead. For the first time, with the experience of the past week behind us, the question was raised whether we had enough time that summer to complete the trip. Already we were two days behind schedule. Perhaps the biggest worry was the long portage that led into the Grand Canyon of the Liard. This was Devil's Portage, seven kilometres long and running over a mountain ridge 400 metres high. Time was becoming a factor. We had to be in Fort Simpson by the 25th of August to catch the chartered DC3 that would fly us back to Fort Nelson. From Hell's Gate, at the lower end of the Grand Canyon of the Liard, we still had more than 500 kilometres ahead of us. Even if things went well and we did not run into any problems in the canyons ahead, we would have some difficulties in keeping up with the schedule.

After a long discussion a decision was made. We would paddle down to the head of the portage, then walk across it, with only a sleeping bag, rain gear and some food. We would then also hike into the canyon as far as we could get in a day to get an idea of what lay ahead. We knew that in doing this we would lose a further two days, we also knew that it could save us a lot of grief later on; for here we entered unknown territory. Not many had come this way since Fred and Charles Camsell had canoed this stretch in the summer of 1899, or if they had, had left no account of their travels.

The next day dawned bright and clear, the sun a golden disk in a blue sky. Shortly after daybreak we slipped into the calm but fast current, the air soft and cool against our faces. The forest mirrored in the water reflected the first flush of blazing fall colours. Four kilometres below the bridge we passed the mouth of the Trout River, its waters the tint of a mountain stream, green, and laden with glacial silt, cold and rushing in foaming waves and riffles.

Here the Liard plunged through the northernmost slopes of

the Rockies. Its valley was broad; the steep, treeless mountains perhaps 1,500 metres high stood far back from the river. We held the canoes close together. Everybody was quiet and nervous. Ahead of us was one of the most dangerous stretches in the river—Devil's Gorge. We could not afford to miss the beginning of this portage. According to the map the river made a tight U-turn here and, over a distance of thirteen kilometres, plunged over continuous rapids, roared over small falls, and twisted itself through gorges whose walls were no more than thirty metres apart. The portage cut across the bend of the river while climbing over a high ridge.

Even this safe trail had quite a reputation; extremely steep on both sides, it led through dense bush and over rough terrain. McConnell, for one, had taken six days for the portage, and had been forced to abandon his wooden boat. To meet such an emergency he had brought with him a roll of canvas in the shape of a canoe. This he now built, using willows as a frame.

"It was stretched," he noted in his report, "on a stout plank hewn out of a small pine tree. Spruce poles, to which the canvas was firmly sewn, were used as gunwales, and willow withes for ribs, while slips to lay between the ribs and canvas were easily cut."

This long and difficult portage undoubtedly also had a lot to do with the decision of the HBC to abandon the Liard as a fur trade route as soon as an easier route was found from the west. It was this long and devilish portage that presented the most difficult problems to the gold seekers, who at times plunged across it in the very depth of winter, in one and one-half metres of snow and in minus thirty and forty degree temperatures.

There were also some vague stories floating around among the local people we talked to that told of a few people who had tried the canyon instead of the portage. They were tales of drownings, for the most part, and our tight-lipped faces reflected our inner tension as we approached the first of the white water. Our eyes anxiously searched the shore for signs of the old portage trail, but none came in view. Too much time had passed, too many storms had howled across the country. Finally, in a likely looking spot we pulled the canoes up on the bank and looked for the old trail on solid ground.

A half of an hour search along the riverbank in both directions and inland revealed nothing. We decided to head out on our own. We had a compass, knew the general direction and the lay of the land, and had an excellent map. Perhaps we would run into the old trail later when we came to the top of the ridge.

And so we began the long hike.

Each of us carried a small pack with food for two days, an emergency ration, an extra sweater, rain gear and sleeping bag. The rest of the provisions we hung between trees, out of reach, we hoped, of bears and then tied the canoes to trees on top of the bank.

Just below the 400-metre high ridge we noticed the first blaze of the portage trail. Looking along the right-of-way, back toward the river, we realized that its beginning was much closer to the gorge than we had imagined. The trail itself was in tough shape. Thick, springy moss on slippery rocks, steep slopes and dense underbrush made walking difficult. Once again we took turns wielding our axes, clearing new growth and marking the new trail, while we struggled endlessly through vast areas of deadfall. Bringing the canoes over this difficult and steep trail would be a considerable challenge.

We hiked for hours through the silent world of bush, until, finally, far below us, we saw the glittering ribbon of the river reappear. Two more hours brought us to its edge and to a view that was overwhelming and breathtaking.

We had come out of the bush at the edge of a 100-metre deep canyon. Far below the waters of the wild and raging Liard squeezed through the last tight spot of the gorge. I had never seen anything quite like it. The entire width of the river was a boiling mass of waves, eddies and whirlpools, of curling breakers, strong currents and even stronger crosscurrents. Impossibly high above these angry waters, hanging on rocks and on ledges, and piled on rocky outcrops, lay the polished and shattered logs of the flood waters. How the river must have looked when it brought these trees down we could only guess. It took only one look to convince all of us that the tough portage was indeed the best and only way, just as it had convinced McConnell and his men and all those who had come before us and had survived.

Below the narrow gorge the Liard spilled into a large bay. At the bottom of it a long, flat sandy beach reached far into the now peaceful waters. It was covered with long grass, a few willows and a stand of young dark spruce trees. This became our campspot, the only sign of life was the deep tracks of a large moose and the larger footprints of a very large and heavy grizzly.

We were exhausted. The day had been a bad one for Jeff. Somewhere along the descent of the slope he had gone off into the bush following a call of nature, and promptly stepped into a hornets' nest. Frantically waving his arms and yelling at the top of his voice, barely holding up his pants, he stormed off, wildly galloping downhill. But the hornets caught up with him and within seconds had done their damage with several stings to his face and neck. The swellings were almost instant until his eyes were mere slits. The laughing sympathy he got did nothing to ease his pain. Dale, Frank and Doug stayed in camp with Jeff while the rest of us started to hike into the canyon. It began a short distance downstream, where the river turned to the left and the high rocky walls once again closed in.

The almost vertical rock faces soon forced us to climb to the top of the bank and to make our way through the bush high above the boiling waters. The hike was tough and yet fascinating. After an hour or more of climbing over dozens of fallen trees and detouring deep ravines with small creeks that rushed toward the river, we reached a high plateau. Here, judging by the size of the trees, a huge forest fire had raged out of control 10 or 15 years ago. Since then though, life had returned to the plateau and had turned it into a beautiful, natural parkland with large stands of young jackpines, all of the same height, and willows and luxurious high grass. The plateau stretched for kilometres to a range of treeless, dark green, nameless hills. Beyond these, we knew, was the Alaska Highway—a two-day march away. On the plateau itself small lakes petered out into marshes that were drained by narrow creeks in gullies trickling down to the river. The entire region, as far as the eye could see, was ideal moose habitat; the cool evening breeze carried their pungent scent. As dusk fell we watched four of the dark shapes as they came down to the lakes to drink, standing near the willows, only to disappear again in the darkening shadows of the forest.

But we had come to study the river and to determine whether or not we could successfully challenge its rapids. We ran and slid down a steep shaly slope to the water's edge and then hiked along the shore again until another vertical cliff face barred our way. It was time to turn back to camp.

We now knew that the rapids in this, the first part of the Grand Canyon of the Liard, could be run after inspecting each set from the shore. They followed each other almost continually as far as we had come. How they looked farther downstream we still did not know. However, it was almost certain that we would have to inspect and judge each rapid before running it, for the entire length of the canyon. Anything else would be too risky in this vast wilderness. From now on we could trust only our own judgement; even McConnell's report was of little or no use here. He had come along this stretch of the Liard during the season of high water and had spent many days portaging canoe and outfit along the rugged bank, sometimes far inland to avoid the gullies. It was certain that this canyon would take considerable time, probably more time than we could afford on this trip.

Dusk had turned into pitch black darkness. We stumbled over rocks and windfalls and wondered what the hell we were doing out here this late at night. An hour passed and then another, until finally, far below us, and still a long way off, we saw the flickering light of our campfire. It was now eleven o'clock and we had been on the go since five that morning. Our days began to resemble those of the voyageurs. Jeff, in great pain, and unable to sleep had kept the soup warm.

Wink, as usual, was up first the next morning but even he had some difficulties in appearing cheerful and had more than his usual troubles in getting the rest of the gang up. Only a nagging hunger and the tantalizing, strong aroma of coffee finally drove us to the campfire. It was time to make the decision to leave this wild stretch of the Liard for a future trip.

To continue our journey to Fort Simpson, though, was still possible, by one of two routes. We could try and run the Toad River which joined the Liard just below the Grand Canyon, or we could drive to Fort Nelson and then follow the Fort Nelson River into the Liard. It entered our river about two day's travel below the Toad.

We knew from our maps that the Toad River also ran through an unexplored canyon of some length before it reached the Liard, and that we could easily spend just as much time in that canyon as we could in the Grand Canyon of the Liard. The Fort Nelson River, on the other hand, would present no problems. There were even barges on the Fort Nelson River in summer that took freight up and down the river between Fort Nelson and Fort Simpson and beyond. The Fort Nelson River it would be.

First though, we had to get back to the canoes, then track them upriver, back to the mouth of the Trout River where the Alaska Highway swooped down to the Liard. There we would put up camp and wait, while Adolf and Wink hitchhiked to Watson Lake to get the vehicles. Around noon, after a six-hour hike we were back at the canoes and, wasting little time proceeded to track them the fifteen or so kilometres upstream past the mouth of the Trout River.

With the long rope tied to the bow thwart and one man sitting in the stern to hold the canoe in deep water, tracking began well

Tracking the canoes upstream near the mouth of Trout River, on the Liard.

even though the high bank was not ideal to walk on. But it didn't take long to realize just how tough this particular job of the voyageurs had been when they tracked their canoes up this river. Not a mere fifteen kilometres as we now attempted to do but over more than a thousand kilometres, and they had done it, time and time again, in under fifty days.

I shall always remember that afternoon, stumbling along that river, soaked to the waist, my running shoes ripped by the sharp rocks. We ran along the gravel bank so as not to slide down into the water with each step, teetering over slippery boulders and climbing over giant piles of driftwood, all the time trying not to tangle the long line. I remember crossing the river again and again, swearing at the partner who sat in the stern, stiff with cold but doing nothing.

Frustration though, never lasted long and finally there was the last bend, and there hung the dust cloud of the highway in the evening sky. We dragged the canoes a few metres up the Trout River and then ferried across its gushing current. Exhausted now we paddled up along the eddies of the Liard, to the mouth of a small creek behind a giant pile of driftwood and pulled into shore.

Here we set up camp. The grassy terrace above the river was a comfortable spot and the gravel beach in front of it and stretching down to the water beside the driftwood pile, made a good kitchen. It took only a few minutes for the tea water to boil and the soup to cook. Then Wink and Adolf took their sleeping bags and walked out to the road to thumb a ride to Watson Lake.

For the rest of us came dry clothing—a luxury. Later we put up the tent on the grassy terrace and piled gear and provisions under the tarp beside it. Some of the smoked meat, packed in plastic bags, had become damp and mouldy and needed airing. Between the fire and the river we spread it on a dry log to let the evening sun do the job.

It was time to prepare a more substantial supper and soon the delicious aroma of cooking food awakened a new hunger and pushed exhaustion aside. Maybe it was this smell, together with that of the smoked meat on the log, that brought a bear into camp. Suddenly it stood not six metres away on top of the bank, its massive head and paws making threatening moves that sent cold shivers of fear racing down my back.

CHAPTER 10
Bear in Camp

The beaver are also abundant, and like the moose, appear to have thriven in the absence of their hereditary enemies. Grizzly bears were reported to be especially common on the Devil's Portage, but we did not meet any.

—R. G. McConnell

WE WERE LOAFING AROUND the campfire after supper, washing dishes, when Dale turned toward the tent. Halfway there he drew up sharply, sucked in his breath and said softly: "Hey, look who is here."

We turned and were face-to-face with a large black bear. He stood at the edge of the bank, no more than six metres away. His head gently swayed from side to side, his nose stretched far forward, sniffing for the smoked sausages and bacon that lay on the log between the fire and the river. It stood on the trail where we had cut steps into the hard clay to make climbing up to the tent easier. His body was leaning forward, his front paws at the sharp drop-off and groping along it. There wasn't much doubt to what he had in mind—to slide down into our kitchen and to get at the meat.

The bear had come silently, and silently he now stared at us. For a moment we froze and stared back and then fear-induced adrenaline snapped us into action. Each man looked frantically for

A frequent visitor to the Liard—the black bear.

a weapon: a wooden spoon, a pot, a heavy stick from the woodpile, an axe. At the same time Jeff began banging two pots together, and, screaming loudly, advanced toward the furry intruder. Harvey and Dale and the rest of the gang joined in, making enough of a racket to shake the leaves out of the trees. The bear, already in the process of sliding down the bank, lifted his head, appeared surprised, hesitated a moment, looked us over with his small black eyes and took a few steps backwards. Then, still not very much concerned, he laid down in the tall grass beside the tent, scratched his belly, and with a hard swipe of his huge paw, swatted Doug's pack into the trees.

Now, for the first time I thought of my camera that lay on my sleeping bag in the tent. This was one photo opportunity I could not miss. Slowly, my eyes on the next climbable tree, I slid over to the tent, keeping as far away from the bear as I could and ducked inside. My fingers shook as I first changed the lens and then the film. The sound of the bear's breathing just outside the flimsy nylon wall of the tent did not make it any easier.

The bear looked bored. Having his picture taken did not bother or excite him. He was still scratching and only occasionally did he turn his small eyes in our direction.

It became clear that something dramatic was needed and soon, if we only knew what. It was quite evident that this bear did not have a casual friendship on his mind, for him we were the intruders and it was us who offered this fine-smelling meat on the log.

"Where the hell is the gun?" Harvey now spoke up. He had thought it over and wanted action. Enough of this uninvited dinner guest.

"Still down by the canoe," Doug answered, "I'll get it."

"And where are the shells?"

"There in that pack, in the outside pocket. But move slowly, for Chrisssake . . . ."

As the rest of us watched, Doug crept off for the gun and Harvey looked for the shells.

"There, I feel better already," said Harvey as he slid home the bolt of the rifle, chambering a shell.

And still we stood and waited, the camera ready, the gun at the ready, only the bear did not move.

"Maybe a shot over his head will help," Harvey said as if he were talking to himself. He aimed and put a bullet into the trunk of a small tree just centimetres over the bear's head. The shot sounded like a cannon in the silence of the bush. The bear looked up, hardly blinked and went back to scratching his belly.

We were getting nervous now. The bear seemed to have settled in and looked as if he was trying to make up his mind on the next move.

"I don't think I want to share the camp with this bear," Harvey said. "Yeah, and this guy obviously wants to stay," chuckled Doug.

But the bear now took the initiative. He too, had come to a decision while scratching his belly. You could see the change of mood in his eyes. Now they held anger. He got up, stretched as if he had just come out of a good nap, and slowly stepped toward us. Equally as slow we pulled back, step by step, away from the fire until the river cut us off. But the bear came on, reared on his hind legs, his mouth open, and swayed back and forth. For him we were the intruders, we were in his territory and it was up to us to get out.

Once more we banged the pots together and yelled—the bear was not impressed.

"One more step and you get the bullet," Harvey muttered, lifting the gun and aiming it. He did not want to do this but for the bear the fun was over. He dropped to all fours and ambled toward the meat. At the same time the shot rang out. The bear collapsed and lay still. He was much smaller now. We buried him in a shallow grave.

The incident had left us sad and nervous. For the first time on this trip Harvey's gun lay beside his sleeping bag while he slept and the axe laid beside my jeans that was my pillow. Only Doug was still mumbling about the bear as he sat beside the fire half the night, sewing up his pack where the bear's claws had ripped it.

We could have slept in the next morning. Wink and Adolf were not yet back, but old habits were hard to break. The day began sunny and warm and soon Harvey and I floated down to the mouth of Trout River in an empty canoe, intending to run the last two or three kilometres of this wild mountain stream.

The short run was well worth the effort of carrying the canoe up along its bank. The river was choppy and exhilarating.

Later in the day we washed clothes, cooked and ate huge meals, baked bannock, sat around the fire and told lies. Finally at around eight that night, Wink and Adolf pulled in and backed their vehicles up to the tent, ready to load up. They wanted to move out right away and in a wild and crazy fifteen minutes the tent came down, the food was packed up and the canoes strapped to the roof racks.

We piled into the trucks and headed out, down the Alaska Highway, glad to be on the move again. The highway was busy. Heavy transport trucks came charging around the curves as if they owned the whole road. I liked the river better. Then there were the rabbits. Never before or since have I seen so many rabbits in one place. They were everywhere: in the shadows along the road, flitting across the gravel, running down the tunnel of the headlights and dead beside the road.

At Toad River, a small community located where the river of the same name ran under the highway, we stopped for coffee and pie, hoping to get some information. It was now one in the morning. The tall American who owned the all-night coffee shop, wore a large black western hat and said that he also ran a string of packhorses and guided hunters in the fall.

Yes, he knew the Toad River, he said, he had ridden along its banks many times. Yes, he also knew the deep canyon, through which, he for one, would never venture in a boat. But no, he did not know the first thing about canoes and was the first to admit it ... surely we would not be that crazy and try it.

We continued toward Fort Nelson.

The coming dawn was a light strip in the east as we drove through the sleeping town and parked under the bridge that spanned the Muskwa River. We spread the tarp on the ground, grabbed the sleeping bags and crawled into them, clothes and all. Five minutes later silence returned to the bridge.

Morning came cold and clear and much too soon. The tent fly that Harvey had spread over the sleeping bags was white with the first frost of the early fall. The warmth of the crackling fire felt good and the hot, bubbling porridge soon heated up our insides. Across from the fire, two elderly ladies came crawling out of the back of their station wagon, walked down to the river and combed their hair. Both in their sixties, they were on their way to Fairbanks. We had to admire them; it certainly took spunk to drive that road alone and to camp along the way. We invited them for breakfast.

For the first time since the day we pushed the canoes into the Liard at Watson Lake, nobody was in any hurry, nobody seemed particularly eager to get onto the slow-moving Fort Nelson River that promised only hard work for the next three days. With a current that moved less than two or three kilometres per hour we would have to paddle hard to make any distance. It was noon before we finally parked the vehicles at the airport and then pushed off. Once more the canoes strung out in the current, moving slowly.

The river flowed quiet and gentle, in long lazy bends between its wooded banks. Paddling was hard work as we put the kilometres behind us, hour after hour, backs bent to the task of getting slack muscles back into shape. Despite the late start we still made sixty kilometres before putting up the tent on a sandbank.

Around three that afternoon we had pulled in on another sandbank to boil a pot of tea and to chat with a group of Indians whose river boat was pulled up on the sandbar. They had a small fire going and were roasting a rabbit. They had killed it, they said, a little ways downstream while inspecting a set of bear tracks in the soft sand. Two of the men had gone after the bear and the rest of the group—an elderly man, two women and a bunch of small children—were now waiting for them to catch up.

The young hunters had no luck. Soon they came walking out of the bush, the guns cradled in their arms. Wordlessly they sat down. We offered tea and cookies. They sweetened their tea with heaps of sugar, then they walked down to the river boat, started up the motor and pulled out into the current, heading upriver toward Fort Nelson. They did not look back. We were alone again.

We were up at dawn the next day. We paddled in the slow moving current of the river, around bend after bend, enjoying the warm sun and wishing for the excitement of white water. More than eighty kilometres farther downstream we camped on yet another sandbar. In front of the tent now stood a tripod of strong willows we had put up and on it hung a large chunk of moose meat. In the frying pan sizzled fresh moose liver and the last of our onions. We had fresh meat—and that without having to fire a single shot! We had "salvaged" the moose meat that was hanging on the tripod from four timberwolves that had killed the calf. The story of how we ended up with fresh moose meat was an adventure in itself.

CHAPTER 11

Moosehunt and Rainstorms

In crossing the portage we started several moose, and it may be mentioned here, that the country we have been passing through and as far as Hell Gate, is probably the best moose country in North America. Everywhere we landed fresh tracks in abundance were observed. We killed one at the mouth of Rivière des Vents, and another farther down near the Crow River, and could have shot a number of others if we had so desired. At the Rapids of the Drowned we scared three into the river, but these unfortunately attempted to swim the rapids and were drowned. They were found afterwards some distance below lodged in a drift pile.
—R.G. McConnell

As always on the calmer stretches of the river, the canoes were strung out that early morning as each team was finding its own rhythm and speed. Frank and I were out front, far ahead of the others, enjoying the solitude of the young day. Over the water lay a heavy mist which the sun, after an hour of playing hide-and-seek, began to burn away. Ahead of us the river flowed in a long straight stretch and then bent into a long lazy turn. The left shore was flat and sandy. Here young willows grew to the top of the gentle slope, only to merge there with the heavier bush of spruce, aspen, poplar and birches. The main current flowed along the steep, washed out right-hand bank that rose three and four metres out of the water. Chunks of hard, dry clay kept breaking off as we passed and clumps of trees slid into the current to create light riffles and eddies.

We were paddling in the middle of the river, anticipating the next bend, when, far ahead, where the shore sloped gently to the water, a black shape appeared to trot up and down along the water's edge. A moose, we thought, but it was hard to judge the size of the animal at that distance.

We drifted and looked back to the other canoes. Doug and Harvey who were behind us, had already seen the animal. Instinctively we all swung closer to the high bank, to blend into the landscape so as not to be spotted too soon.

We were still half a kilometre away when a second animal, smaller, but also black, dashed out of the willows along the river. We were too distant to clearly see what it was, and assumed it was a calf.

Slowly we realized that something was not as it should be down there at the water's edge. The "moose" still charged up and down the shoreline, the second shape following the first like a shadow. Could it be that the cow wanted to cross the river and the calf hesitated? We drifted closer.

Four hundred metres away we began to cross the river at a shallow angle that would bring us to the left bank a few metres above the creatures. Now things began to happen. Four more animals came tearing out of the willows, one of them on the long, gangly legs of the moose calf. The other shapes looked more like large dogs . . . no, wolves! Wolves were after the calf. The second animal we had seen from across the river was not a calf, but a large black timber wolf. It had been shadowing the cow moose, holding it down by the water's edge, away from the calf, while the other three chased it through the willows and then down to the river. Now as we watched from less than two hundred metres away, they quickly closed in and pulled the calf down just as it hit the soft sand and began stumbling toward the water. The shallow water and the soft sand had slowed the calf down enough for the wolves to close in and finish their job. A clever ploy and it had worked perfectly.

We had come a few minutes too late to "save" the calf, for now the wolves looked up, saw the canoes and disappeared instantly in the willows. One moment they were there, the next they were gone, as if the earth itself had opened up and swallowed them. Only the calf lay in the wet sand, half its head in the water, silent testimony of the drama that had taken place. Now the cow moose also looked up as we sprinted the last hundred metres across the river, swung around into the willows, appeared again fifty metres downstream, splashed into the water and began swimming for the far shore. The black timber wolf, too, appeared one more time, if only for a moment, on a low, flat open spot, a few metres away, and then disappeared again together with his white and grey companions. It was eerie how quickly the drama had unfolded and how silently.

We ran the canoes onto the sandbank, took up the cameras and walked to the calf. The wolves had done their job well, the calf was dead and the meat was still warm.

"Well, there is fresh meat," Wink grinned, and with the quick, sure knife strokes of an experienced hunter began to butcher the calf. Minutes later the steaming liver lay on a bundle of clean willow shoots, together with a hindquarter. The rest we left. The wolves would be back; they were probably watching us from the cover of the willows.

What had happened seemed cruel but was part of nature's endless cycle of survival. In these wild parts along the Fort Nelson River the balance of nature was still intact.

We pulled into shore a few kilometres downriver and then hung the meat up on the tripod to let it cool. Nobody had wanted to wait any longer for the fresh liver, which was soon sizzling in the frying pan, together with the last of the onions, a bit of flour, some herbs and a shot of vinegar. Later we sliced juicy steaks off the hindquarter and still had enough meat left over for a stew.

We impatiently wait to devour Doug's mouth-watering bannock.

After supper we sat by the fire, lazy and full. The silence of the country was almost tangible. The sun slowly disappeared behind the dark spears of the northern spruce trees and then turned the river into a sea of red, purple and orange. Doug, who had taken over the job of cook, was baking bannock.

Originally the plan had been to bake bannock each evening to be used for lunch the next day. It never worked, for time and time again the golden brown pan bread was devoured hot from the frying pan, heaped with butter and sweet jam. It was simply too delicious to be left for the next day. It also did a lot toward appeasing our cravings for sweets.

Doug was using a recipe that he had carried with him since his army days. He mixed three cups of flour, a teaspoon of sugar, some salt, two teaspoon baking powder, a cup of raisins and a cup of milk powder in a large pot. He than added enough water to make a smooth batter and put the mixture into the frying pan, to be baked over a bed of coals for about twenty minutes.

"Now the most important thing," he never got tired of lecturing, "is the heat of the fire. It must never be too hot, or your bannock won't have time to rise and will just burn as a flat pancake." Then he'd wink, and grin, "Experience, that's the thing, experience."

Doug of course was an expert and his bannocks always were perfect, soft and golden-brown creations. These were then sliced in half, as soon as we could tough them, covered with butter that melted and dripped, heaped with jam, and devoured on the spot with groans of pleasure.

One of the "assistant" cooks was then delegated to stir up another batch, this time to be kept for next day's lunch. Together with a few tea bags, some sugar, a soup or two, a bit of cheese, dried fruit and some dried meat, it was packed in a special bag. This together with the large coffee pot was tucked under the shroud of one of the canoes. Thus we only had to get the fire going to come up with tea or lunch and did not have to dig into the food boxes until it was again time to camp.

The stew we had prepared though, with the last of our moose meat, did not turn out well at all. To give it more substance, we had, along with the usual ingredients, added a couple of handful of barley. We left the heavy pot simmering on a bed of coal most of the night and had even gotten up around midnight to put an extra log on the fire. In the morning when I lifted the lid, a horrible smell almost threw me back. Somehow the barley had begun to ferment during the night and our stew was now covered with a thick, smelly layer of grey foam. Holding my nose I took a stick, grabbed the pot and dumped it among the willows on top of the bank. Perhaps some coyote would have use for it.

By mid-afternoon that day—it was now the middle of August—we once again reached our old friend, the Liard. If it had not been for the stronger current and the cleaner water, we would have hardly noticed the confluence of the two rivers. Even though our maps indicated two cabins and a floatplane dock with the name "Nelson Forks" written underneath, we saw no signs of civilization. Both the Liard and the Fort Nelson were spread out over many channels, separated by large and small islands. Obviously we had picked the wrong channel and had bypassed the cabins nestled behind some island. But that was all right with me—I could do without people for a few more days.

Instead we met more bears. Coming around a bend in the river we spotted a large black bear swimming for the opposite shore. Intending to get a few close-up pictures I sprinted after it, while Frank, sitting in the bow, started to dig the camera out of his waterproof bag. The bear was still a hundred or so metres from shore when I gently nudged his backside with the bow of the canoe and then stayed with him stroke for stroke. I was shaking with laughter, but Frank did not think it all that funny. He glanced nervously over his shoulder a few times, wondering just what the hell I was doing, and frantically began to clamber out of his seat and back over the thwart and the shroud to get away. At the same time he was trying to operate the camera.

The poor bear was frantic, fairly ploughing through the water, the bow of the canoe firmly at his backside, Frank huffing and puffing with irritation. Then the bear touched bottom. By this time he was so upset that he didn't know whether he was still swimming or already running, tried to do both at once, and bounded through the water throwing spray into the air. With a full head of steam he flew up the bank and crashed into the willows. He did not even take time to shake off the water and I could still hear him crashing through the woods when I turned the canoe, while Frank, still mumbling to himself slipped back into his seat.

We put up camp on a high, level, grassy terrace above the river. At the back of the small clearing stood three dilapidated cabins half-hidden in the thick willows, berry bushes and tall grass. They were wintering cabins of native hunters and trappers. On their walls hung homemade snowshoes and the frames for stretching beaver and otter pelts. Packrats had taken over, left their mess and smell and we soon decided that we preferred to sleep under the stars. Perhaps that was the wrong decision, for the relative comfort of the long, soft grass was soon upset by the attack of hordes of mosquitoes. Apparently they too, had found the long grass an ideal shelter, and for the first time since we had left Watson Lake we were really bothered by these stinging devils.

We took it easy the next day and enjoyed the mild, early fall weather. Every hour we drifted for a while, letting the slow current do all the work. We laid back in the warm sun, to roast our already deeply-tanned skins. Cabins now appeared on both sides of the river, together with many tent camps, a sure sign that we were getting closer to the next settlement—Fort Liard, one of the earliest Hudson's Bay posts in this region. In and around the cabins and tents, kids were playing but as soon as they saw us, they stopped and stared. If we only made so much as a gesture to approach them and the shore, they quickly disappeared together with the few women and they went behind the cabins or hid in the tents. Perhaps because the men were away hunting and fishing, they did not seem to want anything to do with these strangers who came paddling down this river. Once or twice we tried to talk to the people but were met with stony silence. We drifted on.

Only once did that change, farther downstream, where a camp stood close to the river bank, in a small clearing where the whole

An Indian family smokes moose and bear meat on the shores of the Liard.

Indian family was at home. Beside the tent, on racks of willow poles, hung strips of bear and moose meat over a smoking fire. Perhaps it was this assured supply of winter meat that made this particular family friendlier than the others. Whatever the reason, they waved and gestured for us to land on the sandy beach below the camp, where the Indian had pulled up his river boat and where an old Chestnut canoe was tied to a tree.

Taking cameras we scrambled up the steep bank. The smell of half-smoked meat was overpowering and the acrid smoke of the green wood fire soon had our eyes streaming.

The man, tall and clad in wool pants, moccasins and a checkered wool shirt, on his head an old cap, waited at the edge of the bank and shook our hads. Though he spoke little English beyond yes and no, he never stopped talking in his native tongue, while we all smiled foolishly and tried to figure out just what he was saying. He had the typical broad face of the northern Indian and slightly slanted eyes. He had also lost all his teeth. Somewhere between thirty- and sixty-years-old he was obviously happy with his life, and for some reason still more happy to see us. He grabbed a piece of freshly smoked and blackened meat—the shoulder blade of a moose—and handed it to us as a gift. Laughing I accepted the meat and then went down to the canoes to get a couple of cans of butter with which to return his generosity. He flooded us with words we could not make out and shook hands all around again. Even his small, round wife, who up until then had remained in the tent, now smiled a toothless grin and stepped out into the open to lean against the meat rack, not minding the smoke at all. Only the two young girls, perhaps ten and twelve years old, were still much too shy to show themselves fully and only looked at us from behind the tent, with beautiful black eyes and black, stringy hair hanging into their faces.

A half-a-dozen tall, skinny dogs knew no such shyness and probably would have torn our heads off had not their ropes tied to young poplar trees held them back. They tried to make up for it with gnashing teeth, snarling and frantic barking, until well-aimed chunks of wood and a few sharp words from the Indian drove them behind the tent and into whimpering silence.

The camp of this native hunter was in a good spot. The old Yukon tent stood in a small clearing overlooking a long stretch of the river. Inside stood a wood stove and a pile of blankets made up the beds. Beside the tent, and beside the smoking fire, on yet another rack, hung the hides of two large moose and the heavier skin of a large black bear.

Later in the day we tried to eat the half-smoked meat but the smell of the half-cured meat, leathery and black, was too much for our stomachs. And so we threw it to a hungry-looking dog that had appeared as we stopped for tea and now stood well back to stare at us. His sides were gaunted, his hair was dirty and matted and his watery eyes never left us. We had no idea where he had come from but he seemed to like the meat which he grabbed as soon as we pulled out, to trot up the bank with it.

Around four in the afternoon we pulled across the mouth of the Petitot River and then pointed the canoes toward the white, square building of the Hudson's Bay store at Fort Liard.

The small settlement with its white painted houses, its old Hudson's Bay store, its new log school and its weathered log cabins and tarpaper shacks looked almost prosperous. At the foot of a long, wooden stairway the RCMP riverboat with its twin outboards was tied up. A tall, native constable was tinkering with one of the engines. Below it, and also tied up to the floating dock was a green forestry boat and beside it, a small red plane on floats. Soon we sat on the dock to change shoes and shirts in an effort to look halfway respectable before going to the store. I don't think we succeeded. A three-week beard and the generally greasy and grubby appearance acquired in the close proximity of countless smoky fires in the bush could not easily be hidden.

The hanging wooden steps led up the bank to the main road of the settlement. This was a dusty gravel strip, running the length of the village and parallel with the river, only to stop at the edge of the bush. Strung out along it and burning in broad daylight were street lamps. Beyond the road were green, well-kept lawns, white board fences, lush and blooming flowerbeds and the houses of the few whites—a teacher, a nurse, the Hudson's Bay manager and the

RCMP corporal and his wife. Beside the houses were vegetable gardens with corn and potatoes and cabbages. Between the school and the river were a large playing field, a flagpole and a fancy carved sign that said: Fort Liard Pop. 276. The people were making the best of their isolated location.

We turned toward the Hudson's Bay store that stood, whitewashed and trimmed with red, by itself on the banks of the Petitot River. We were hoping to replenish our supplies of sugar, jam and perhaps get a few fresh fruits. But we were disappointed. The young manager only came to the door, opened it a crack and told us that it was Saturday, that he was doing inventory and that the store was closed. We could not convince him to let us in. Disappointed and more than a little angry at the stubbornness of the clerk we turned away. We would just have to make do until we reached Fort Simpson.

Later we wandered over to the police station to report on our progress and were promptly invited for tea and cake. During the conversation with the corporal and his wife we mentioned the plight with the Hudson's Bay store, and unexpectedly it brought results. The friendly officer, dressed in uniform shirt and blue jeans, said he was sure the clerk would make an exception and then walked back to the store with us. And so we got our needed supplies, despite the bad mood of the clerk, and then headed out once more toward Fort Simpson.

Later in the evening hovering clouds, heavy with water, burst into a steady drizzle. When I awoke around midnight a heavy rain was drumming steadily onto the fly of the tent.

It poured all night. The sleeping bags were damp and those that had pressed too tightly against the walls of the tent were wet. Cranky and still tired we got up early. Anything seemed better than to lay in the damp tent and to listen to the pounding of the rain. Luckily it stopped just long enough to make breakfast and to load the canoes. Shortly before six we squeezed into the seats, tightened the shroud and pushed off. Paddling hard we tried to get warm.

The sky hung low and dark and cold over the dripping forest. It started to rain again, and only a narrow strip of clear, blue sky in the west let us hope for better weather. But, when an hour later a strong wind came up, blowing straight into our faces, and pushing heavy squalls toward us, we knew we were in for a wet day.

The face of the Liard had changed dramatically. Its waters now rolled in dark, strong, brown waves toward the north. After two or three hours we were wet through, despite our rain jackets. The steadily increasing wind whipped the rain horizontally, and beat into our faces like small hail. The waves, too, grew steadily higher. The storm, blowing against the current, dug the river—here wide and flat—into a fury. It transformed the once quiet and placid waters into a sea of breaking waves, foaming whitecaps and deep, shifting troughs.

We paddled on, chins tucked into our chests, faces dripping with water, arms aching from trying to hold the canoe on course, from paddling against the relentless wind. Again and again we tried to find shelter in the lee of the high bank, were forced to cross the river whenever it swept around the long, open bends. It was almost as if this were some sort of test we had to take, that the gods of nature had flung at us to prove our mettle. The day turned into a long and constant battle with the forces of the cosmos, against the violence of the elements.

We tried to head into the wind, fighting against it, our canoes rolling heavily as they were lifted up, again and again by the steep waves and then slid down into the next trough. We tried to stay close together, in case of emergency, but it was difficult in the strong wind and the high waves and we were afraid of crashing into one another. Soon each canoe was on its own until we tried to group up again and to stay in sight of one another.

It was a wild, fascinating and crazy day. The wind, the waves, the canoes seemingly always far from shelter, forever battling against the furious forces, struggling in the endless sea of angry white waves and dark, deep troughs. Amongst all this, wildly blowing ponchos; glistening, dripping paddles; tucked in heads with eyes half closed against the driven rain; dark squalls with grey, fast-moving clouds that seemed to have swallowed the trees of the dripping forest.

Toward noon, my arms now two cold, hurting stumps that threatened to fall off, Harvey and Doug finally headed for a sandy beach. Pulling the canoes up onto the sand and securing it to a tree, they sprinted into the shelter of the trees for a lunch break.

It was a bad place with flat, level land, soaked in water, dripping birches and poplars bending in the strong wind and driving rain. But we landed the canoes, hands cold and almost numb, shivering uncontrollably in the first stages of hypothermia, and followed them into the trees. Desperately we tried to start a fire in the damp forest. Kneeling, swearing, teeth chattering, we huddled in front of the little pile of mostly damp sticks to watch again and again as the tiny, flickering flame died into a wind-whipped wisp of smoke. Nowhere, it seemed, was there a single dry piece of wood. Only when Wink finally ferreted a few pieces of dry birchbark did we manage to coax reluctant embers into a blazing fire.

Half an hour later things had improved considerably. Between the trees roared two large fires, while eight men, soaked to the skin, huddled close to the flames, and then stood, turning around and around to dry their steaming clothes. With the warmth, hunger also returned. A thick soup was brewed—nothing ever tasted better.

The heavy rain was still slicing horizontally in screaming squalls across water and land and the dark, heavy clouds still whipped through the tree tops when a couple of hours later we returned to the canoes. To continue on was out of the question yet we had to find a better campsite. What we needed were large spruce trees, with sheltering branches that would break the wind and keep off most of the rain. Quickly we tucked away the tea pot and soup kettle and slid into the canoes.

Wink and Jeff pushed off first and wasted no time crossing the river to get into the lee of the high bank. Fifteen minutes later we spied what we were looking for—a small grove of tall spruce trees, high on the bank.

An hour later we were almost comfortable. We sat under the tarp, the gear stacked away, the canoes high on the bank and secured with ropes, a fire blazing. With the axe we had cut steps into the clay bank, while a tight rope for a handhold had helped move gear and supplies up the steep bank.

Getting camp ready though had been a wet affair. The first task was the fire; already some of us were shivering again. We also cut poles for the tent and for an additional shelter we put up a large orange tarp over the kitchen. Then we had to get enough firewood to keep the fire going for most of the coming night. It is hard to imagine that you can get wetter when you are already wet to the skin, but gathering firewood (dry standing trees are best), in dripping willows, as dense as a jungle will do it. Soon our clothes were glued to our skin.

We spent the rest of the afternoon in the relative comfort of the camp, roasting in front of the fire, lying in the sleeping bags, cooking and eating huge meals. When Doug had the brilliant idea of breaking open the emergency ration of overproof rum, our spirits began to brighten considerably.

It was still raining the next morning when we decamped. But the wind had died down and activity seemed preferable to staying in the damp tent. Loading the canoe was a hard and dirty job. Sliding down the steep, slippery bank with gear and provisions was only the beginning. Sinking up to the knees in the soft gooey muck along the shore while loading the canoe was worse.

We stayed close together that morning, which was indeed fortunate. As we followed the high bank of the river, a large she-bear almost jumped into Adolf's canoe!

We probably would have missed her if she had not grunted from the top of the embarkment as we floated around an outside bend, close to the high cutbank. Her fur was jet black and dripping wet. She was nervously stalking back and forth along the top of the bank rearing half onto her hind legs at each change of direction, her front paws grabbing air, nearly losing her balance at each turn. For a moment we wondered what she was doing, but soon the reason became clear.

Two young cubs, no bigger than coyotes, tried to climb the slippery bank, whimpering. Apparently they had all swum the river and now the two cubs were stuck at the bottom of the steep, high incline. We back-paddled and watched. Again and again the young cubs tried to scramble up the slippery slope, only to lose their footing each time in the top quarter and to slide back down, squealing with fright and frustration. We drifted closer and began to yell encouraging words at the cubs. Mother bear did not like our intrusion and became still more agitated. Two or three times she threatened to fall into the river or into Adolf's canoe which was directly below her. In the meantime Dale had taken to slapping the flat side of his paddle onto the water, to encourage the cubs into still more frantic efforts. With loud grunting and squealing—answered by mama bear with a loud roar—they panicked, threw themselves against the clay bank and scrambled up, legs flying. For a moment they seemed to hang in the air, just below the lip of the bank, and then made it over the top. There, as if they had not suffered enough, their mother unceremoniously cuffed them into the bushes and disappeared on the run.

Early in the afternoon we began to look around for a campsite but it was almost five o'clock before we finally found what we wanted. A lone spruce tree, high on the bank in a grassy clearing, commanded a marvellous view over the valley of the Liard. The rain had stopped around noon and by the time we slipped into the sleeping bags, the wind had sprung around and began to clear away the last of the rain-drenched clouds.

I woke up about two in the morning under a canopy of bright stars. I quietly strolled down to the river's now gently flowing waters. There was a cold solitude as only the north can produce.

To the west a pale slice of the autumn moon hung behind the towering spears of the spruce trees, bathing the land in its eery, cold light.

Then the northern lights began to flicker along the horizon and I knew what had awakened me at this unusual hour. The strange green curtain of light that danced back and forth along the sharp outline of the horizon shot long vertical strips into the sky, undulating with incredible speed. They bundled together into bizarre forms, disappeared and reappeared, and then slashed again across the heavens in long vertical stripes, more yellow than green and almost crackling.

I don't know how long I stood there in the cold silence, overwhelmed by the spectacle of nature. Perhaps an hour. Then, chilled to the bone, I climbed back up the trail and slipped into my bag.

The new day dawned sunny and then turned mild. The storm had passed like yesterday's memories. The country was washed and clean, and paddling became a pleasure again. The Liard River was now spreading out, split into channels by large islands, some of these stacked with huge piles of driftwood.

We had already seen islands like these that are forever changing the face of the river before. Begun by layer upon layer of driftwood that the high water deposited, they were sometimes three and four metres high, the logs still clearly visible where the strong current washed against them. Later the river had brought sand and earth. Willow shoots had started to grow. Many years later, other trees began to grow on that island which changed the course of the river. The single stages of this development were clearly visible and proving that these changes are ongoing. Where the bank fell off abruptly the trunks of the driftwood logs were bound together by a maze of roots. On the other side, where the shore climbed gently out of the water, the different stages of willows could also be seen. Down by the edge of the water the willow shoots were tender and green and only half a metre high.

A watched pot never boils! Lunch break on the Liard, with the thick, dense willow stand in the background.

Behind them stood another row, twice or three times as high, and farther back, clearly defined and still taller stood another. Here the willows were thick and then seemed to melt into the dense bush of the region.

CHAPTER 12

Nahanni Butte

In the first of these bends we met with a crew of Hudson's Bay voyageurs in charge of W. Lépine, who were endeavouring to make their way upriver to the mouth of the Dease. Lépine had been employed on the river as a guide in the old days, when goods were taken by this route to the Yukon, and was well acquainted with it. He brought news of scarcity of provisions in the Mackenzie District, and this decided me to send my two men back up the river with him, and to depend on the services of natives for canoemen in the future. Lépine had become disheartened by the continued high water and the difficulties of upstream navigation and when we met him talked of returning, but we induced him to persevere. A small spruce bark canoe which an Indian and his wife built in an afternoon, in addition to the large birch canoe which they already possessed, furnished sufficient accommodation for his increased party, and on the 28th July, after a day's delay, he proceeded upriver. I afterwards learned that with the exception of an upset, caused by the unskilfulness of Trépanier, one of my men, the journey was successfully accomplished and Dease River reached in safety.

—R. G. McConnell

For two or three days now we had been paddling through a flat and endless country with forests stretching as far as the eye could see. Then on a warm and sunny day we once again began to

Wink Bradford and Jeff Harbottle in front of Nahanni Butte near junction of Nahanni and Liard Rivers.

meet the mountains. Naked, rugged ranges of grey-brown rock, without foothills, climbed out of the bush like the backdrop to a stage. They were part of the Liard Range which, in turn, was part of the Mackenzie Mountains that run for hundreds of kilometres to the north. One peak, more prominent than all the others, and much closer, was especially impressive. This was Nahanni Butte and it guarded the junction of the South Nahanni with the Liard.

Nahanni Butte stayed in front of us for hours, tantalizing, seemingly always just around the next bend in the river, but never coming closer. As the Liard slowly made its way towards the Mackenzie, it sometimes almost doubled back on itself, laying its crooked course in giant oxbows, and growing wide and slow. We paddled on the broad concourse for hours and counted ourselves lucky that the storm of three days ago had not hit us on this stretch; here we would have been much more vulnerable and would have been forced to stop and seek shelter in the trees.

Around noon a wind came up. This time, though, the forces of nature blew into our backs. Quickly we rafted the four canoes together and hoisted a large tent fly on the spare paddles for a sail. For five or ten minutes the canoes road majestically like some ancient ships, while we leaned back in our seats and congratulated ourselves on our quick thinking and ingenuity. Ah, this was the real stuff, letting the river and the wind do all the work while we relaxed and sucked on some candy.

Then the wind died. The makeshift sail hung limp. Humbled, we went back to paddling.

Nahanni Butte still loomed ahead, now dominating the landscape. Shortly after noon we came to the first side channel of the Nahanni River where it spilled its waters into the Liard. Before this the Liard current had slowed down as the waters of the Nahanni—dark and muddy—began pushing against it. We turned into the side channel to paddle up it for three or four kilometres, to get to the settlement that was named for the mountain that loomed over it.

The Nahanni was running high after the three day rain in the mountains. Its normally clear and placid waters at this point, near the junction, now rolled in heavy, dark brown waves and at great speed towards its still larger brother, the Liard. In many places the river had topped its bank, roared through thick stands of willows along its edge, while carrying dozens of uprooted trees in the grasp of the main current. The roots of the trees were like giant wheels with clumps of earth and rocks still attached. The Nahanni looked impressive, and dangerous but there was no other route. We had to work our way upstream as best we could, for we had planned to stay over for a day and climb Nahanni Butte.

At first it didn't go too badly. The current in the side channel was relatively tame and it was possible to make slow progress with the paddle by staying close to the trees and working the eddies. Then where the channel hit the main current things looked bad. Again there was no choice, we would have to ferry across the main channel—here two hundred metres wide—to get into yet another back channel. Once there we would have to work our way up it, cross the main current again, to finally reach the settlement. We tried to work our way up the main channel but soon found it impossible to make any headway. The water was racing through the trees and there were no eddies here. After ten or twelve metres we gave up, peeled off and got set for the long ferry.

It seemed to take hours. Setting the canoe in a steep angle into the current and fixing a point on the opposite shore we paddled hard so as not to lose too much ground. Luckily, by now we were in shape. With quick, powerful strokes and changing sides every ten strokes or so, we drove the canoes into the current. Even to hold it in one spot, to let a giant tree drift by, we had to paddle all out, only to redouble our efforts when the next clear stretch opened up. Slowly the far shore moved closer, although it seemed that we were forever locked into these high, rolling, dark waves, arms and shoulders sore, lungs gasping for breath.

Then finally we were in a large eddy at the foot of an island that split the Nahanni. We grabbed some willows, held on, exhausted.

It did not take long to recuperate. After a few minutes we worked our way up the narrow side channel, to the top of the island, and there, across the river, was the settlement—newly painted log cabins, a red plane on floats down by the wharf, up on the bank a small church and a brand new school. Beyond that was the long rectangle of Nahanni Butte's airstrip, cleared out of the bush.

To reach it we had to cross the river again which was now close to four hundred metres wide. Again we pointed the canoes into the current, waited a few seconds to let some trees drift by and pushed off. In midstream we hit a sandbank, that, in lower water was a flat island. A few large evergreens, hung up on the sand, gave us a chance for a much needed rest. Then slowly, sinking almost knee-deep in the soft sand, we dragged the canoes upstream as far as we could, all the time watching for the drifting logs that seemed to be more numerous by the minute. When the sand disappeared beneath us, we jumped into the canoes and fought the raging current with hard, short, choppy strokes.

On the opposite shore, up on the high bank above the parked plane, the local population was gathering. We hardly noticed them. We were too much involved with the river, with holding the canoes in the current at the right angle, so as not to be whipped around, or worse, caught broadside. We were too busy dodging the floating trees and sprinting in the clear stretches between them. In the slow water along the bank we turned downstream and drifted to the dock.

We tied the canoes to the dock beside the plane and talked to the pilot who was filling the gas tank. The heavy rains had flooded out a mine some hundred kilometres upstream, and he was busy ferrying the seventy or so workers out to Fort Simpson.

Later in the afternoon we crossed the Nahanni once more, higher up this time, to put up camp at the foot of the 1,800 metre high mountain that had guided us for such a long time. We erected the tent at the edge of the winter road, which, the map indicated, ran all the way to Fort Simpson. These so-called roads, roads in name only, are put down in winter by bulldozers guided by compass and allow snowmobile travel between remote settlements through a country that in the summer is impassable muskeg. The creek, though, which we had expected to find in the draw, had dried up and only half a kilometre farther inland we found a good spring. The river water, thick as soup with floating sand, did not look very appetizing.

The landing at the campsite in the high water had been a messy one, even though a large pile of driftwood slowed the current along the shore. Stepping out of the canoe we sank up to our knees in the soft muck of the gently sloping beach, and by the time we had unloaded the canoes, we had trouble getting out of the quicksand-like substance.

Only a good supper and a wash in the lively trickle that was the spring set things right again. Later in the evening we went to visit Dick Turner—trader, trapper, pioneer and recently author of *Nahanni* and *Wings of the North*, who was then still living in his large log house across the river from the settlement.

Dick had come into the country some forty years ago. He had recently given up the trading post he had run for years. Down by the side channel of the Nahanni, in the calm water of a large bay, his aircraft was tied to a dock. He still flew it, despite his age. In the yard stood a yellow snowmobile—it had replaced the dog team on his trap line—idle now in the late summer.

Trapping and the general fur trade had not been too good lately, Dick explained. But the box of manuscripts in his hands when he opened the door proved that he had found other things to do. Dick and his wife talked of the pioneering days that evening and did not let us return to camp until we had admired their beautiful vegetable garden and had accepted enough fresh carrots and potatoes to last the rest of the trip.

We slept in the next morning and it was after seven when we finally finished breakfast. Five of us then packed a lunch and headed for Nahanni Butte.

At the beginning the hike was easy as we climbed the first slope and then followed a well-travelled game trail. It led along the edge of a steep rocky wall that plunged some two hundred metres straight down into the heavy bush. Halfway up the mountain a recent landslide had knocked out the trail and forced us to struggle through the dense underbrush until we hit the treeline at about 1,200 metres. From there open alpine country led up to the rocks that marked the highest spot, now adorned by a makeshift pyramid of peeled poles and red plastic streamers. The cairn that marked the summit of Nahanni Butte had been put up by a party who had come by helicopter a few years earlier.

Overwhelmed by the panoramic view of the vast country that stretched below us, we gazed down into a land of forest and water through which we had come and would still go. To the north, the Liard, now united with the dark, dirty waters of the Nahanni,

Adolf Teufele (right), Jeff Harbottle and Wink Bradford (middle). View from Nahanni Butte.

serpentined toward the Mackenzie in countless bends, like a giant snake writhing through endless muskeg. From the west the Nahanni emerged from snow-covered mountains. Calm and wide in its lower regions, it gave no hint of its spectacular canyons and waterfalls (it was declared a National Park years later because of these land formations). To the north and west we looked into the Mackenzie Mountains, peak marching upon peak, an endless country, still largely unexplored, the home of a few Indians and of mountain goats and mountain sheep, of large moose and giant grizzlies. And far below stood the tiny houses of Nahanni Butte, the airstrip a narrow scar in an ocean of forest with more shades of green than you could ever have thought possible.

Squatting behind the rocky cairn, trying to get out of the cold wind that swept across the peak, we made a small fire, cooked a soup and brewed a tea. We did not stay long but packed up to hike across a wide, sweeping saddle that connected with the second peak of the Butte. Later we glissaded down a sharp incline of loose rock and shale at reckless speed, yelling with excitement like a bunch of kids, and only stopping when young poplar trees led into a steep draw. The dry stream bed was soon an impossible tangle of deadfall and large boulders and we were forced into willow thickets that were so dense that we soon lost sight of each other. Farther down the slope we hit a well-travelled trail, which, as a large pile of recent droppings indicated, was the domain of a bear. Fresh tracks also told us that the trail had been travelled that very day. That knowledge and the imagined snorting and grunting of a large animal at our back, did not make for a leisurely Sunday stroll and I was glad when we hit the winter road that led us back to camp.

After dusk had settled, a heavy storm came up that soon whipped through the trees around us. I was lying in my sleeping bag writing up my notes by flashlight, when a loud groaning and screeching noise high up in a tree just outside the tent made me look up. Harvey, too, had heard it and together we slipped out of the tent. In the beam of the flashlight we saw that we had put up the tent under a dead spruce that leaned against another large spruce. The eerie noise came from the rubbing-together of the two trees, whipped back and forth by the strong wind. We tried to judge where the tree would crash in case it slipped off, and for a while considered moving the tent. In the end, though we were too tired to bother and slipped back into the tent. I tried to sleep but without much success, wondering just when and where the tree would crash to the ground. Only towards midnight when the wind died down, did I fall into an uneasy sleep, while telling myself that in future I would also look up when next putting up the tent in the bush.

We left Nahanni Butte early the next morning and were moving downstream shortly after six. Before us now were stretches of the Liard where water and sky merged. Stretches also, where the river turned back on itself in giant loops, and stretches where, for each kilometre that brought us closer to the Mackenzie as the crow flies, we paddled three in loops and bends. All day Nahanni Butte hovered at our back, until finally, late in the evening, the mountain disappeared behind the straggly spruce of yet another river bend.

We put up camp at the mouth of Birch River and swam in its dark water, amongst large, smooth boulders that were the remnants of the glaciers of the last ice age. It was not easy to find a good place for the tent and we worked for more than an hour to clear enough willows to make a level spot.

Two hours into the next morning's paddle we hit the last of the whitewater, a stretch of about twenty kilometres of rapids, which, according to Dick Turner, could all be run along the right-hand bank. It felt good to get into white water again, as we worked our way down from chute to chute, from eddy to eddy. When we reached the last of the obstacles, the Beaver Dam—a rocky ledge that stretched across the entire width of the river—we pulled into a quiet bay. We unloaded two of the canoes, tracked them back upstream to run the rapids in the middle of the river, where the waves were highest, just for the sheer joy of it.

I shall never forget the height of those waves. I still remember talking to myself, as Adolf and I approached them, "They can't be this high, they can't be this high . . .!" But they were, and as we cut through them with too much speed, they were over my head and crashed into the bow, half filling the canoe. The waves and the water sloshing in the canoe fought us every metre of the way back to shore.

Below the Beaver Dam the Liard grew wide and lazy. We drifted most of the afternoon, laying back on the covers, enjoying the warmth of the sun, reluctant to bring this trip to an end.

It had been getting tougher and tougher to find a good campsite for the past few days. The whole country was now flat muskeg and that evening, our last in the wilderness, was no exception.

The next day we reached Fort Simpson at nine in the morning, two hours after we had pulled out of camp. We had meant to sleep in, but the rocks and roots that poked at our bodies through our down-filled sleeping bags like hard, accusatory fingers, did not give us any peace, so we packed up.

CHAPTER 13

The End of the Trip

Fort Simpson is the oldest post continuously occupied on Mackenzie River. The site was first occupied by the North West Company in 1804. The Hudson's Bay Company took it over a few years later about the time of the amalgamation of the two companies. The post faces Mackenzie River on an island just below the junction of the Mackenzie with the Liard. The river is here almost exactly a mile wide with a strong smooth flowing current. One can stand on the bank of the river and look upstream for about 12 miles with the huge promontory of Gros Cap in the middle distance separating the Mackenzie from the Liard. Below, the river runs straight away for many miles. The site is the grandest on the whole river but its location on an island is somewhat of a handicap, for the island is cut off from the western mainland during periods of high water. Also, with the opening of the river in the spring, it stands in some danger of being overridden by the ice when a jam develops in the river below.

—Charles Camsell, Son of the North (1954)

T<small>HE WIDE SANDY BEACH AT</small> Fort Simpson, and the high, sparsely vegetated bank behind it, didn't look very inviting on that early morning.

Fort Simpson, once the headquarters of the Mackenzie District of the HBC, was, even in the early 1970s typical of the loneliness and remoteness of the northern bush settlement. The houses were small, the unkempt front yards covered in tall weeds. Facing the wide and dusty main road that ran down the middle of the settlement, stood a new post office, a Hudson's Bay store, a garage and a police station. The community hall where the weekly dance took place on Saturday night was also on the main drag. Set back from the road were the log houses and tarpaper shacks of the local Indians. Facing the river, instead of the road, were the old buildings of the former trading post, arranged, as McConnell and Camsell had described them, in a large square.

It was a Saturday morning when we arrived. The streets were empty. The village seemed dead as summer's grass as we walked along the main road, sporting our four-week-old beards and greasy jeans but our presence had not gone unnoticed. A tall, good-looking Indian stepped out of one of the houses on a side street. Holding him by the hand was his five- or six-year-old daughter. He stopped in front of us, swayed slightly from side to side, offered to shake hands and with a heavy tongue introduced himself as John Baptist, crane operator. Only after we later talked to some of the local people did it become clear why the man was swaying as he walked this early in the morning. Each Friday the government's liquor store was opened, the only time during the week, and a supply of beer and hard stuff sold that was flown in on a DC3.

The HBC store opened just as we arrived. We were soon crowding around the meagre display of fresh fruit. It didn't matter that the prices for apples and pears were astronomical because we had been in the bush for more than four weeks and had not had a chance to spend any money along the route. We felt like kids at a fair.

It felt strange to suddenly see people again, and dirty, mud-spattered, snarling pickup trucks. After buying a few things we returned to the canoes to make lunch—already we missed the tall spruce trees. We decided to rent a flatdeck truck and to transport our gear and canoes to the airport, some fifteen kilometres out of town, and to camp there. City life was not for us.

THE END OF THE TRIP

We set up camp on the soft grass beside the runway, to wait for the arrival of our plane but the joy of a good campsite was short-lived. The airport manager came out to announce that we most definitely could not camp within the airport's boundaries. So we once again returned to the willows. There was no more doubt, we had arrived among people again—their silly rules and regulations were proof enough.

Later in the afternoon we went back into town to take in the weekly dance. "Come on in," the taxi driver had said. "It's quite the thing." And it was.

I have never seen such panic that took hold of the people when the doors of the hall were thrown open at 7.30 p.m. and the bar started to do business. Three and four rows deep the men and women, natives and whites, crowded up to the bar. All waved their ten dollar bills, and all wanted to be served first. Here the beer was not sold by the bottle, but by the case. In ten minutes or less the mad rush was over. Everyone—man, women, native and white, was seated at the long tables with one or two dozen bottles in front of them, and a more serious business of drinking I have never seen. There was only one object—to empty these bottles as quickly as possible. The ritual was fulfilled with a determined single-mindedness and matter-of-factness that astonished me. In minutes the tables were littered with bottles in all stages of destruction.

Even the band could not rouse the drinkers at the start of the evening. Only when everyone was slightly tipsy did the people begin to stomp to the beat of the music but they never forgot their sole objective—to get roaring drunk. The beer was sold out a half hour after the bar opened up and the sweating bartenders began to serve the hard stuff, whisky and rum mostly, at a dollar a shot. An hour before midnight even that supply was gone. And the dance was getting ugly. It was time to head back to camp.

"What kind of bar is this," Doug kept saying, "sold out before midnight . . . eh, what kind of bar is that?"

There were sore heads even in our camp the next morning

when we finally arose to spread another tarp over the tent. It was raining hard. There was not much else to do but stay in our sleeping bags, suffer our hangovers and write up notes.

In the late afternoon the rain slackened, and towards six we heard the noise of an approaching plane. Harvey, who had been visiting the airport manager, came running into our camp, "all right you mountain bums, the plane leaves in fifteen minutes."

New life surged into tired bodies. Within minutes our camp was packed up, our bundles and canoes loaded onto the DC3, the plane took off, our trip was finished.

Far below the Liard disappeared as the plane banked steeply and then set course for Fort Nelson where we would unite with our vehicles. Below us were the wide bends of the Fort Nelson River. We had paddled it with tired muscles, seemingly a long time ago. There was the beach where we had watched the wolves kill the moose calf. Then it too was gone.

The Grand Canyon of the Liard, though, was still waiting.

CHAPTER 14
Grand Canyon Days

The story originated with a Yukon trapper and prospector by the name of Tom Smith, who, with his daughter Jane, came to the hotsprings from the west in 1923. Smith and his daughter, who was only fourteen or fifteen years of age, lived at the hotsprings for two years and must have been quite successful in trapping furs.

With their two year's catch of furs they set out on the river for Fort Liard. They traversed the Devil's Portage, ran the Rapids of the Drowned successfully and navigated the whole canyon. With all the dangerous water of the Liard behind them they were able to relax. Through some mishap, however, their boat was swamped in a riffle, their whole outfit lost, and Tom Smith drowned. The girl hung on to the upturned boat and was swept onto a sandbar. Shortly afterwards she was rescued by a band of Indians who took care of her and ultimately turned her over to the Hudson's Bay Company at Fort Liard. Later in the summer she was sent to the Anglican Mission at Hay River on Great Slave Lake and there she died in 1934.

—Charles Camsell, Son of the North (1954).

T·wo YEARS HAD PASSED. A trip into another part of the world was behind me—into the hot plains of the Serengeti and to the cold peak of Mount Kilimanjaro—before I again saw the suspension bridge that crosses the Liard River near the Hotsprings.

We were back to try the Grand Canyon of the Liard, four from the old team—Adolf, Harvey, Wink and I—and two new faces, Mike Stein from Vancouver, and Jim McConkey who ran a ski school in the Coast Mountains of British Columbia. The Liard had been on my mind often during those two years, especially the Grand Canyon. According to all descriptions it was not only by far the most beautiful stretch of the Liard, but also the most exciting and most dangerous. Charles Camsell in his book *Son of the North* wrote:

> *I have travelled many thousand miles in a canoe, and it is a pleasant way of travelling, particularly downstream but I know of no more enjoyable canoe trip than the one from Dease Lake to Mackenzie River. Few travellers have ever canoed the whole length of the Liard River and fewer still have attempted its passage in high water. Of these, only R.B. McConnell of the Geological Survey of Canada, and David Hanbury have left a detailed record of such a trip. When we descended the river in August, 1899, in the season of low water, it was a relatively wild stream. Yet there was sufficient hazard and excitement to make the journey stick in my mind when the experiences of other exploratory trips have faded out.*

Our second trip down the Liard River began at the Alaska Highway bridge, at Kilometre 794 of the Alaska Highway. Ahead were ten or twelve kilometres of relatively calm waters. I remembered them well. Here we had tracked the canoes upstream last time but the water was much lower now and it took only two hours of steady paddling to reach the head of Devil's Portage.

Something, though, was changed and for a while we could not figure out just what it was. Suddenly it hit me, there had been a forest fire, a roaring blaze that had burned both sides of the river for many kilometres. The marker we had left, a spruce tree with all the branches trimmed off except for the crown, in the tradition of the voyageurs, was gone. And gone, too, was the trail we had slashed out.

After putting up the tent on the riverbank, Harvey, Wink and I set out to see whether we could find traces of the old trail, perhaps farther inland, after all, we had blazed it clearly only two years before. The difficulties began less than a hundred metres away from the river. Trees—mostly northern spruce—half burnt, lay on the ground by the hundreds. Just as many were half fallen and together they created an almost impassable black tangle of upturned roots, charred trunks and branches. For a while we tried to hack our way through the mess, but soon we were forced to skirt along the worst spots parallel to the river. An hour passed, then another. Faces and hands were black, clothes torn. Only when we began to climb steeply away from the river did we get out of the fresh burn, but by this time we had wandered much too far downriver. From the top of the ridge we could see the Liard far below begin its bend around the Devil's Gorge. Across the river we could make out the steep draw where Deer River gushed down at the very top of the bend. We could hear the thunder of the rapids even at this distance, while the foaming whitewater among drenched black rocks, was clearly visible. At that spot, according to our maps, the river was barely forty metres wide, and below the mouth of the Deer River the canyon walls closed in to within a mere thirty metres. Nothing had changed—the long portage was still the only sensible access to the Grand Canyon of the Liard, but to get canoes, provisions and gear across it, through this fresh burn, with its almost impassable tangle of downed timber, would be as difficult a task as we had ever attempted.

We turned back and reached camp just before dark. The rest of the gang had supper ready. Over steaks and fresh potatoes and numerous cups of hot tea, we discussed the plans. To look further for the old trail was impracticable. Instead we would chart a new course across the ridge as best as we could, hack out a new trail and pack canoes, gear and supplies in short stages as we went along.

The long and difficult portage around Devil's Gorge.

First daylight found us well under way inland, packing heavy loads, sweating and swearing. Harvey and Wink were up ahead somewhere, clearing trail with their axes, wishing they had brought saws. Most of the windfall was too heavy to be chopped with the light axes we carried and we were soon forced to climb over, or, just as often, slip under the half-burned trees. Progress was slow and tedious. The personal packs were not too bad, but the heavy food boxes were still full and close to fifty kilograms (a hundred pounds) each. The five-metre canoes often were almost too much. Time after time the canoes had to be lifted over the fallen trees and each time they grew heavier.

At noon we cooked a soup. We looked as if we had been there a week, faces black with soot and ashes, and smeared with sweat.

Late in the afternoon when we had worked our way about three-quarters of the way up the first slope, Wink discovered a terrace-like depression where a clear spring flowed out of the burnt soil. It made a good campsite. The rest of the slope to the top of the ridge was getting even steeper, and, as a quick scouting trip ahead revealed, the other side was just as steep and flame-scorched. Right now it did not matter; we were too tired to worry about the next day. Completely exhausted we leaned the canoes into forks of trees, put gear and provisions into a neat pile, cooked a quick meal and went to sleep.

Daylight came only too soon. The hard work of cutting trail and lugging canoes and gear up the mountain continued. Only occasionally were there any green patches left, small areas where the fire, for some reason had missed a few trees while all around the destruction was almost complete. The day passed slowly, but then the steep, four-hundred-metre-high ridge was finally behind us. We had been forced to cut steps into the last section and had started to descend the other side, knees shaking under the heavy loads. Again it was late afternoon and the Liard was still hidden behind endless bush.

We needed water. There was only one thing to do: to put canoes and gear into a pile—hoping that no grizzly would find the food—take a small pack with sleeping bag, tent and some food, and search for the elusive river.

Halfway down the slippery slope we finally came out of the burn and the going was much easier. At nightfall we slid down the riverbank to the water's edge. On top of the high embarkment we put up the tent and turned in.

Early the next morning we hiked back, muscles stiff and sore, and found gear and canoes just as we had left them. It was noon again before we had moved everything to the river.

Seven kilometres in two-and-a-half days—but it was done. Once more the canoes were where they should be, on a gravel bar by the river. Here, after its wild romp through the narrow canyon it flowed almost smooth. A swim in the cold water of the bay brought new life into the tired limbs of the paddlers. In the afternoon we basked in the warm autumn sun.

The tent stood in a small clearing forty metres above the fast moving water. Around it grew young spruce trees and willows and

High shale rock formations are part of the Grand Canyon of the Liard.

in front of it, beyond a narrow, grassy strip, the gravel and clay bank dropped almost straight down to the water. To the south, the northernmost peaks of the Rockies carried the first dusting of winter snow. To the east, the last gap of the Devil's Gorge opened up into a large bay below us. In the canyon upstream the Liard boiled and roared. Below it the valley was wide and peaceful. At its lower end was the wide, sandy strip of the peninsula where we had camped two years before. Harvey and I pushed a canoe into the water and paddled down to it. Nothing had changed since we had been there last. Only the tracks of moose and bear were fresh in the sand. The fire pit, the tent poles we had leaned against the trunk of a tree, were there as if we had left them only a few days ago. People did not often traverse this pristine wilderness.

An extra day in camp brought us much needed rest but we were not idle. We hiked upstream, along the top of the canyon and looked down into the Devil's Pool, the river a turmoil of eddies, whirlpools and giant boils. Farther upstream the canyon walls narrowed still more and the water there was a frothing, frantic surf. On rocky points and ledges, ten, fifteen metres above the present water level, hung the bleached and polished tree trunks of the flood waters. We were glad to have picked the land route.

The following day dawned clear and cold. We were now well into Indian summer with its cold, frosty nights and lovely, mild days. The aspens and birches of this northern forest already wore fall colours and stood in brilliant yellow along the edges of the river. Only the willows in the steep ravines, where we had fought our canoes down to the water's edge, were still green.

I had spent a good part of the rest day re-reading McConnell's report. In it he mentioned an account of two men, who were said to have descended the entire length of the Grand Canyon of the Liard in a little over two hours, all fifty kilometres of it! An incredible feat—if true. McConnell himself had expressed his doubts. He and his helper had struggled for six days just to come across the portage, and then spent many more long days in and out of the canyon, making tough portages forced upon them by the high water. But the first morning as we were underway again, and less than an hour into the canyon, we were almost willing to believe the story of these unknown miners. Here were indeed currents that rushed by the sharp and jagged rocks at twenty kilometres an hour and more.

For us though, there was no reason to hurry, or to take chances. Safety was the all-important thing, and so we landed again and again, secured the canoes in shallow bays and calm eddies and walked ahead to inspect each rapid as we came to it. We were simply too remote from any kind of outside help and could not risk losing a canoe, or worse, a life.

The rapids we encountered, scouted and then ran that first day in the Grand Canyon of the Liard, were typical of that stretch of the river. On both sides sheer, rocky, shaly cliffs climbed almost vertical out of the water, the black shale up to a hundred metres

Unloading the canoes at Liard Spire, the tall rocky tower seen in the background.

high. From time to time, jagged, rocky ribs, sticking far into the water, caused large breakers to run diagonally to the main current. Where these diagonals met in foaming V's, in the middle of the stream, were the safe chutes. Most of the time there was not much room to manoeuvre, since we tried to stay just off-centre of the large standing waves at the bottom of the chutes, to avoid swamping the canoes. Behind the rocky ribs were the eddies we used for stopping places, from where we scouted the next portion of the river. Most of the eddies were strong and violent, requiring perfect timing, and we always tried to cut across the first diagonal to catch the very top of the eddy. Long ago we had learned that the middle part or the bottom end were usually too turbulent for our loaded canoes.

The morning soon turned into a glorious white water run. At noon we pulled the canoes onto the sand behind Surrender Island in a spot that was ideal for a lunch break. Soft, untouched, white sand sloped gently to the water. Farther back began a narrow strip of grass, where late flowers were still in bloom. And beyond that was bush, low cranberry bushes, dense underbrush, then spruce and poplars and birches. Where the sand met the grass we built the fire and then settled back to enjoy the warm sun.

This was the good life—mild Indian summer weather, an incredible, azure blue sky, downstream more dark, shaly rocks and more challenging, boiling white water.

We came a lot farther that day than we had anticipated—close to twenty-five kilometres. Shortly after three we began to look for a campsite. We knew that three or four kilometres farther downstream were the notorious Rapids of the Drowned and we wanted to inspect them on foot, with plenty of time to pull out above them. From the right Sulphur Creek came out of a long, deep cut; the smell of sulphur from the mineral springs along its banks hung heavy on the water. Past it on the left, a sheer, rocky tower, twenty metres high, reached into the sky. The tower was a landmark, the remnant of a once solid band of hard rock, that had run along the entire section of the river. We named it Liard Spire. Beside it a small creek that came out of a deep, rocky gully joined the Liard. The creek had formed a small, sandy delta, overgrown with willows and, on a small perfect flat, sparse grass. It was our camp spot. With quick, strong strokes we ferried across the rocky mouth of the creek, turned into the eddy below and jumped into the water to lift the canoes onto chunks of driftwood to protect them from the sharp rocks that lurked just below the surface of the water.

An hour later the tent was up; on tight ropes hung wet socks and sleeping bags both destined to soak up the evening sun that slanted across the river. After tea we slipped the cameras from their watertight bags and walked downstream for a firsthand look at the Rapids of the Drowned.

McConnell had mentioned the rapids in his account and explained how the got their name:

> *At the end of this reach it bends to the north, and striking violently against some sombre cliffs which line the left bank, is deflected again to the east with the formation of what are known as the Rapids of the Drowned. Here, one of the most dangerous spots of the river is formed by the water plunging with its whole force, over a ledge of rocks which curves outwards and downwards from the left bank into a boiling chaudière behind. The name of the rapids originated from the drowning at this point of a Hudson's Bay clerk named Brown, and a boat load of voyageurs. As the story goes, Brown, disregarding the advice of his steersman insisted on running close to the northern bank, and the canoe plunging into the hole mentioned above was drawn under.*

"Luckily we have low water," I said, as we walked along the shore toward the infamous spot, "and nobody along named Brown!" For the first few hundred metres we followed the shoreline, walking along sandy beaches and climbing over large boulders which, long ago, had broken off the canyon walls and crashed down to the water's edge. The walls of the canyon began to close in, became steeper and higher until, finally, they fell straight into the water. Climbing along them became dangerous; we were forced to scramble to the top of the hundred-metre-high bank and to inch our way along its edge.

Wink Bradford (bow) and Harvey Fraser (stern) coming through standing waves in "Rapids of the Drowned."

The ascent was tricky. The near vertical rock face offered few good handholds and the churning eddies far below did nothing to ease the mind. Only halfway up were we able to climb into a deep cut where a few scraggly spruce trees and willows had managed to

Canoeing high waves of the Rapids of the Drowned, fearing its appetite for boats.

take hold. Even at the top of the bank, that stretched away in a high plateau, the going was difficult. Dense underbrush and the ever present tangle of windfall made walking difficult, while steep gullies falling down to the river's gushing waters forced us to make long detours inland. There were no signs that anybody had walked along here since McConnell almost a century ago had struggled past this point with boat and gear. There were no blazed trees and no old trails, only when we finally came out onto a well-travelled moose trail that twisted steeply down to the river over a gentle hogback, did things look up. We slid down the trail and then climbed downriver over rocky bands where the river closed to within fifty metres.

Below us the rapids boiled. The keeper behind McConnell's rock with its chaudière or vertical eddy, was unmistakable.

At first glance the passage looked indeed difficult and risky. There seemed to be no clear path for the canoes. Would this be another long and difficult portage? McConnell, we knew, had lined his boat on ropes here. He had then climbed out of the canyon, only to be forced into a long and treacherous portage that had gone over many kilometres. Because of the steepness of the canyon walls he had been unable to get back down to the river. Of course, he had come through here in high water—the river was much lower now.

Slowly, as we studied the rapids, trying to read the currents, a route became clear. True, it would be a fairly difficult one and would require some precise manoeuvring, but with a calculated risk the rapids could be run. Halfway between the middle of the river and the left bank, a clear chute ran between two rocks, it looked safe enough. Below the two rocks the canoes had to be shifted hard to the right, probably best done with a back ferry, to miss yet another rock and a strong keeper behind it. Immediately below that the canoes had to be shifted to the left again if we wanted to catch the first eddy below the rapids where a good sandy beach offered a safe landing. Below the rapids where the river narrowed even more, a half-metre-wide rocky ledge running at an angle out of the water to a height of some five metres, offered the ideal spot to set up Adolf's film camera to record the event. The rapids looked exciting. As long as we could remember the route, we thought optimistically, we would have no troubles.

The next morning, crouching on this rocky ledge, the camera ready, I eagerly waited for the other two canoes to come through the rapids, and for the sun to fall into the canyon to allow picture taking. Adolf sat on a second, higher ledge, his film camera on a tripod. We had come safely through the canyon in the early morning to set up our cameras.

Half a kilometre away, suddenly the other two canoes came bucking around the last bend. In regular intervals they disappeared behind the waves, only to reappear on the crest of the next wave, the bow of each craft climbing high out of the water and then slamming down in a shower of spray. Once again the old rule had proved true—the waves were a lot higher when seen from the canoe than when looked at from the top of the riverbank. Only the tight spray covers repelled most of the water. Harvey and Wink were in the lead. They had no problems staying on course despite the high waves, as they powered their canoe toward the chute, slipped between the rocks and then back-ferried to the left and past the keeper. Holding my telephoto lens on a tall stick for stability, I shot frame after frame, until Wink's face, grinning and excited, filled the viewfinder. Then the canoe whipped into the eddy below me, followed by Mike and Jim, both yelling with the sheer joy of an adrenalin-pumping run.

That day turned into one of the best on the whole trip. We felt free! Around us was pristine wilderness, a wild, beautiful and undisturbed nature. We had found a harmony with the river.

We had learned to adjust to its moods. We were no longer fighting it; instead we felt privileged to travel its spirited waters. We felt as if the Grand Canyon of the Liard had accepted us, and was willing to tolerate our passing.

I had felt the mood already the evening before, back at the camp at the foot of the Liard Spire. We were sitting around the cooking fire and watching the sun slip behind the edge of the canyon. Twilight settled into dusk, then into a velvet-black moonless sky studded with stars. Then, on the opposite shore, a cow moose stepped out of the willows together with her high-legged calf. Side by side they stepped down to the water for a drink. We could hear their slurping in the stillness of the evening. We watched them disappear as quickly as they had appeared. The light breeze that carried the smoke from our cooking fire and the smell of man upstream and away from the cow and calf, had prevented them from noticing us.

The prospectors who had come upriver through this canyon, on their way to the goldfields, had rarely known such peace and harmony with the Liard. Where we now leaned against some rock on soft sand, more than seventy years later, eating jerky and drinking sweet tea, those men had struggled for their lives in numbing cold. On the trail already for endless weeks, hundreds of

The mighty Liard holds the secret of this boiler which has given name to Boiler Canyon.

kilometres into the wilderness, they felt their first signs of scurvy. Doubts about the wisdom of their undertaking had long overshadowed their dream of riches. Now they were forced to concentrate on survival, on sticking it out, day after day, until they could reach the next Hudson's Bay post.

My thoughts were still on these men when we packed up lunch and then pushed off again. Now the Liard ran fast but smooth. Beyond the next bend it would boil again.

About four kilometres below the Rapids of the Drowned, in a spot the map called Boiler Canyon, the dark mouth of a cave appeared on the high water mark on the left bank. In front of it, some three metres above the present water level, lay polished driftwood. On the roof of the cave, amongst some tall grass, I glimpsed what looked like the teeth of an old cast iron gear. Surprised and not quite believing what I had seen, I yelled at the others to pull in. Quickly we landed on a sandbank, pulled the canoes up and then climbed the rocky bank to the top of the cave.

There in front of us lay the wreck of a small paddle-wheeler. Scattered over a few square metres were the remains of a small steamer: the long, heavy driveshaft of steel, complete with the flanges to which the wooden paddles had been fastened; gears, half a metre in diameter; water and steam pipes; taps of copper and brass; a large monkey wrench; oil drilling equipment as it was used at the turn of the century, including a half-a-metre-long drilling bit, shaped like a huge cold chisel and a hundred metres downstream, wedged into the rocks, and half filled with them, was the old rusty boiler.

How this small steamer got this far into the canyon is still a mystery. Even to this day, after some intensive research and a dozen letters, including to the manufacturer of the drilling equipment in Petrolia, Ontario, the Liard still guards the story of this boat that came upstream into the canyon where even canoes had troubles going down.

We assumed that the steamer came upriver in high water and had then been smashed up on top of the cave by the heavy turbulences it would have caused in high water. Here its journey had come to an end. Nobody seems to know what happened to the crew or even when all this took place. Who were these crazy men who brought the little steamer into the canyon? What happened to them—the river guards their secret. We took pictures and a few token souvenirs and then moved on. But we now knew why the river in this stretch was called Boiler Canyon. It was not, as we had assumed, because of its boiling waters, but for the little boiler of the wrecked steamer.

Ahead of us were the Gates. We had marked this spot in red on our maps and expected difficulties. But we were lucky on that first trip, thanks, mainly, to the low water. What we found was a beautiful stretch of white water, with clear chutes and safe eddies where we could catch our breath and laugh with the sheer joy of it.

We put up camp a short ways below the Gates in a large bay and close to a small creek that spilled out of a deep ravine. It had

Looking upstream to Hell's Gate.

started to rain, lightly at first, but by the time the soup came to a boil, heavy drops hammered onto the plastic sheet we had spread over the kitchen and the tent. In front of it, the downpour hit the dark waters of the river. It had changed its colour drastically, from a fresh sunlit green, to a dark forbidding grey. It flowed fast but

Two canoes doing the "Liard-Waltz" through Hell's Gate—downstream end of the Grand Canyon of the Liard. (1972 Expedition).

without riffles. On the opposite shore a sheer cliff of almost white sandstone rose straight into the dark heavy clouds that now hugged the canyon walls. We could hear the faint rumbling of yet another set of rapids beyond the next bend, despite the heavy downpour. We would tackle them in the morning. According to our maps it was only some six kilometres before we came to Hell's Gate, "the end of dangerous navigation," as R.G. McConnell had marked the place on the map he had published with his report in 1887.

Around midnight the rain stopped. By mid-morning the sun broke through the clouds. A steady breeze began to clear away the wispy fog from the hillside that bordered the Liard. We packed up and pulled out. An hour later we landed at the mouth of a nameless creek. Its delta, three hundred metres wide, and strewn with rocks and broken trees, belied the peacefulness of the three small trickles that made up the creek, and spoke of more violent floods. Half a kilometre below, stark, grey rock walls climbed straight out of the river and seemed to close it off completely.

There was no doubt; we had reached Hell's Gate.

We pulled the canoes up on the beach beside the creek and then walked downstream along the shore towards the famous gate. A rocky island, a hundred or more metres high, and covered with a heavy stand of spruce, split the river. But only the right hand channel (it is no more than thirty metres wide) carried water. The left channel seemed to be blocked off by a band of rocks, worn smooth by the water's action. Just upstream from this, and studded with half a dozen boulders, was a quiet bay, a good place to land. We ferried across the river, landed and pulled the canoes onto the polished rocks. We wanted to climb the island in the gate to photograph the canoes as they went through the last gap of the Grand Canyon of the Liard.

Climbing the rocky wall to the top of the island was not much of a problem. Half a dozen scraggly spruce trees and low bushes that grew in the cracks provided good handholds. The view from the island was magnificent.

Upstream, the narrow valley of the Liard through which we had just come stretched as far as we could see. The rapids that had provided such pleasure, and fear, were now reduced to vague white spots in the clean, green water. Narrow side-gorges washed out by the small creeks, which from the river had looked like nothing more than short ravines, in fact ran far back from the river. They cut up the land into a spruce covered patchwork of ridges and valleys and hogbacks. In the distance we could still make out the last peaks of the Rockies, nameless, round, green hills jutting into a perfect azure blue sky. Above the far-off mountains white clouds billowed. Closer to the river the poplar and birch trees were a golden, fiery yellow against the dark green of the spruce that lined the horizon like a row of spears. Far back, the highest peaks were clad in the first snow of the coming winter.

Downriver, the Liard spilled into a large bay, closed in on three sides by vertical rock walls and then disappeared around a long, sweeping bend, moving steadily north toward the Mackenzie. The

From the top of a cliff, Hell's Gate looks almost peaceful.

highwater channel behind the island was tight and narrow. Here the floodwaters had cut their way through solid rock. Three dark green and incredible clear pools, lined with rocks, were the only remains of the spring torrents.

While the rest of the group climbed back down to the canoes

Adolf and I sat on a boulder to photograph them. In a few minutes they came into view. They were like toys, so small, dwarfed by the height and performing, one last time, the Liard Waltz. Hell's Gate, however, posed no problem in low water and soon the canoes curved out of sight, to pull into a sandbar where we had agreed to stop for lunch.

Half an hour later Adolf and I too were through the gate and pulled into shore to look back. Halfway down the wide bay, where we had a fire going, the rocky walls seemed to completely close off the river. There was nothing to reveal where it was coming from, as if it wanted to keep its most beautiful valley a secret a while longer.

The beauty and the challenge of the Liard River though stayed in all our minds. We knew that we would be back again, to experience once more the river that ran still wild and free.

Two years later, when Harvey and I came through the canyon again, this time guiding two Americans on a twenty-five day trip, we camped on the opposite shore just upstream from Hell's Gate at the end of a particular wild day, in a whole series of wild days.

The Liard was running high that year, up to three and four metres higher than in any of our previous encounters with the river. Many sections and large stretches of the Liard were, in fact, almost unrecognizable. Most of the rapids were far more violent in high water, including the Gates. In order to run the river at all we had stayed rafted together through most of the Grand Canyon and still had hair-raising moments. Such as the one in a spot we called "Boulder Bar."

Boulder Bar was a high, gravelly bar, or low island, strewn with head-sized rocks, that split the river into two channels just upstream from where we were now camped. In low water the passage past the gravel island had been easy, despite some high standing waves where the two channels came together again. In high water though, the rapids looked formidable. Boulder Bar was almost totally under water, only a narrow strip, dark and wet with spray, at the very centre showed above the rushing current. We pulled in on the island and dragged the canoes onto some large rocks, to walk to the bottom end of the island, to get a firsthand look at the river below it.

Downstream and to the left where that channel boiled toward the middle of the river and then bent into a left-hand turn, the water flowed through a tight gap and was solid white. Giant breakers rolled off two rocky outcrops that stuck far into the stream. Spray in the form of rooster tails shot three metres into the air. We knew that below each of these outcrops were violent eddies, more like whirlpools than mere backwaters, and they would be waiting. An upset in any one of these was almost certain, a passage on that side impossible.

Straight below Boulder Bar, where the two channels merged, the standing waves we had run so easily in low water were now gigantic—two metres high and more. They rolled away, one behind the other, as far as we could see, where the river curved out of sight. Studying them we knew that this time they were simply too high to handle, even with the spray covers and the two canoes lashed together.

That left the right-hand channel. It too rushed by in heavy rolling waves and with incredible speed. Even beside the bar the waves were high and then washed into the standing waves below. But almost straight across from where we stood, a rocky ridge jutted into the river. Below it a large eddy looked smooth enough if we could hit it at the very top. To get to it, though, would be tricky. Only if we ferried back upstream as far as we could over to the right-hand bank, and then peeled off quickly, could we perhaps get into the smooth slip that washed past the ridge and into the eddy below. That way we would miss the large standing waves and cutting across the first diagonal that came off the ridge, we would be in the eddy. There was little margin for error. Much of the channel washed toward the rocky outcrop of the right-hand side from where it was then thrown violently back.

Judging from the distance we would not have much more than half a metre on either side of the canoes. Just looking at the spot got the adrenaline pumping.

We dragged the canoes, still rafted together, back up the bar as far as we could. Then, taking a deep breath, we pointed them upstream to ferry over toward the right-hand bank. Our two American friends, Bill and George, were excited and did not quite realize the danger. The two were professionals from California and had, up to booking a canoe trip with us, experienced many of the white water rivers of the American Northwest in large rubber rafts. Throughout the trip we had a tough time convincing them that running white water in a large rubber raft was totally different from running white water in a loaded canoe. We always felt that they were rather disappointed when we tried to avoid huge standing waves that would have swamped our canoes; standing waves that in a large raft would have merely been a wild ride.

Harvey and I were nervous. Paddling with hard fast strokes we realized instantly that the current was much stronger and faster than we had anticipated. We were in danger of being carried into the standing waves below while still facing upstream, or worse, broadside while peeling off. We abandoned the ferry, peeled off instantly and drove the canoes hard forward, across the channel, aiming the bows for the smooth slip beside the rocky outcrop. Still, the canoes were being swept downstream much too fast.

"Pry," yelled Harvey at Bill who was in the bow of his canoe. And Bill pried, as we had taught him over the past few days. Perhaps we had taught him too well—the paddle blade snapped with a loud crack in the force of the river. He grabbed the spare behind him, fastened to the top of the shroud with a strong rubber band. Again he pried and crack went the second paddle. Bill looked amazed.

"Chrissake, take it easy," I yelled over the roar of the water and handed him the second spare from my canoe. By now we were shooting straight for the rocks. It was my turn to pry the bows of the canoes back to the left. I planted the paddle alongside the gunnel, beside the knees, and pulled on the shaft. The wooden paddle was alive in my hands, I could feel the blade quiver and the shaft bend. Instantly I eased up, changed the angle of the blade. Now Harvey on the opposite side in the stern was also prying. The two crafts straightened out in the current, shot past the rocky wall with maybe twenty or thirty centimetres to spare, the angle of the canoes pointing toward the eddy. I leaned far out to the right, reached across the diagonal wave that was washing off the rocky point and planted the paddle in the quiet water of the eddy in a high brace and hung on. Again I could feel the wooden paddle quiver. I was almost pulled out of my seat. The force of the current racing along the rocky wall dragged the canoes in a violent arc, pivoting them around my paddle until we were pointed upstream and swung into the eddy. Quickly we pulled into shore and jumped out. Breathing hard, faces flushed with excitement we looked at the broken paddle shafts that Bill had kept on the canoe cover. Now we could laugh. We had come into the eddy where we had wanted to, even if it had been a close call.

We had learnt to work with the Liard, not against it. The respectful bond we had forged with this magnificent, spirited river was exhilarating.

Paddling downstream for three days to Fort Liard and later flying out to Fort Nelson, was anticlimactic. The best part of the river, steeped in history of the fur trade, of the gold rush, of McConnell's exploration, was already a memory. We could only hope that this relatively little known river with its magnificent Grand Canyon of the Liard would be able to keep its wildness and would one day not be lost forever—drowned in one more stagnant lake, behind yet another dam.

CHAPTER 15

Doomed to be Dammed?

Engineering and environmental overview studies which were completed in 1978 concluded, on the basis of preliminary data, that hydroelectric development of the Liard River appeared to be technically feasible and economically attractive. Those studies indicated that the potential power benefits of the development are great and that, while there would be environmental and socioeconomic impacts, these would be within manageable limits.
—B.C. Hydro, Public Affairs, *Introduction to Progress Report Liard River Hydroelectric Development (July 1981)*

THE LIARD, OF COURSE, HAS NOT succeeded in hiding its Grand Canyon, nor its other magnificent rapids and its wild beauty from the world. Most unfortunately, not from that group of men and women who see rushing waters merely as a source of more electric power to be harnessed; who see narrow gorges and canyons merely as perfect places for more dams.

Surveyors, dam engineers and hydroelectric specialists, who view the beauty of wild water in turbines and not in rapids, have all been to the Liard. Preliminary studies have been conducted at to major potential dam sites—at Devil's Gorge, and the head of the Grand Canyon of the Liard, and at Beavercrow, just upstream from the junction of the Beaver River with the Liard, some fifty kilometres upstream from the mouth of the Fort Nelson River.

Maps have been drawn and redrawn that show the extent of man-made flooding—not the possible route for a portage or the passage of a whitewater canoe.

What puzzles and intrigues me is how anyone who knows the full magnitude of these potential projects can still say that the environmental impacts are within "manageable limits," as is stated in the introduction to the progress report. How can an impact be within manageable limits, when that impact kills an entire river; when its concourse would be wiped off the face of the earth and turned into a giant, 400 kilometre-long, stagnant, stump-infested lake; when wild rapids are flooded; when a fifty kilometre-long canyon is filled to the brim?

The power potential of the Liard, of course, is huge, as are the profits to be made from the sale of such power.

The good news is that, for the moment, the Liard is still safe. In a letter of December 15, 1986, Mr. D.G. McFarlane, then manager, System Planning Division for B.C. Hydro, wrote the following:

> ... the Liard River studies were suspended indefinitely in late 1982 due to a reduction in the forecast growth of electric energy demand in British Columbia. ... The Liard River is not being considered for development within B.C. Hydro's current 20-year planning period. A number of potential energy resources, including the Liard, could be considered to meet domestic requirements beyond the 20-year period. However, we have no plans at the present time to resume the Liard project studies.

The 20-year planning period is still in effect, but the power potential of this river must be extremely tempting for B.C. Hydro. The two-project scheme called Devil's Gorge—Beavercrow, that was preferred at the time of the studies, was a scheme that could provide approximately 4,800 megawatts of generating capacity (one megawatt is 1,000 kilowatts or 1,000,000 watts), and 26,000 gigawatt-hours of average annual energy (one gigawatt-hour is one million kilowatt-hours). Make no mistake, this is an incredible amount of energy. The average energy output of the Liard development would in fact be twice that of the G.M. Shrum Generating Plant, on the Peace River.

But at what cost to the river? The Devil's Gorge project with its 200-metre high, earth-filled dam, would create a reservoir with a surface area of 870 square kilometres at the normal maximum elevation of 572 metres above sea level. This may sound innocent enough, but when examined more closely it reveals some rather frightening factors. Not only would all of the rapids of the Liard be drowned, such as Portage Brulé, Whirlpool, Mountain Portage, Cranberry, Little Canyon, but numerous other rivers and streams would be affected as well.

The lower half of the Kechika River up to the junction with the Turnagain River. The Coal River, flooded as far north as the BC-Yukon boundary. The Smith River, also flooded to the boundary. The Trout River, backed up to where it meets the Alaska Highway. The Rivière des Vents, and all the other streams and creeks. All their valleys would be flooded; all the moose and bear habitat would be drowned; all the habitat of timber wolves and all other living creatures. All the vegetation, groves of birches, poplars, spruces and willows would be lost.

The second potential dam site, Beavercrow, with its 160 meter-high earthfilled dam, would create a lake with a surface area of 190 square kilometres and extend to the dam at Devil's Gorge with a maximum elevation of 390 metres above sea level. This lake would put underwater all of the Grand Canyon of the Liard with its Surrender Island, Rapids of the Drowned, Boiler Canyon, and Hell's Gate.

Gone would be the trails of Lee and Camsell and their fellow Klondikers. Gone the portages of McConnell and Dawson and the trails of the voyageurs. Gone the paths of Robert Campbell and McLeod.

Throughout the progress report of B.C. Hydro, there is not a single mention of the beauty of this magnificent river which would be annihilated, and not a word of the history of the river. It is as if the tremendous undertakings of such courageous men as the Hudson's Bay voyageurs, the Klondikers, McConnell, Camsell, had never taken place.

It is true, though, for the moment the Liard seems safe. But a lot of money has been spent in these preliminary studies. As a whole I am more optimistic now for the future of the Liard than I was ten years ago. There is no talk now of damming the Liard River. But if things go as they have in the past, the men and women who view the beauty of a rapid only in terms of the power a turbine puts out, might still have their way. If not now then later.

Right now the good intentions for the Liard River may prevail. But, as William James once said: "With mere good intentions, hell is proverbially paved."

Glossary

Blaze Axe mark on tree, made by chipping off strip of bark.

Bowman The bow is the front of the canoe—paddling in the bow.

Chute Smooth water between rocks.

Ferry Crossing the river at an angle to the current either forward or backward.

Keeper Vertical eddy behind rock which can pin a canoe or person.

Outcrop Rockface over river.

Oxbow Wide turn in river, often dry old riverbed.

Peel off Term for canoe coming out of an eddy into main current of the river, facing upstream and then "peeling off" to face downstream.

Riffle Light rapids in fast water.

Rollers Big waves usually coming off outcrops or boulders.

Shroud Canoe cover.

Spray cover See shroud.

Thwart Cross piece in canoe between the gunnels to hold shape or beam of canoe.

Windlass Mechanical device to pull up or lower load on rope or cable by winding rope onto drum or roller.

Index

Alaska 11, 103 (Fairbanks); see also Alaska Highway
Alaska Highway 3, 10, 26-27, 82, 88, 94-95, 102, 136, 161
Americans 50, 102, 155-156; see also Alaska; California; Chicago; New York; Maine; Missouri; Oregon; Washington
Anderson, James x, 37, 45-46
Arctic Ocean 39
Athabasca District 45
Athabasca River 54, 72
Atkinson, Mr. 77

Balaam, Mr. 81
Bannock (recipe) 109
Baptist, John 132
Barker, William 84
BC Hydro 159-160, 162
Bears 30, 87, 92-93, 97-101, 103, 110-112, 114, 118-119, 128-129, 139, 161
Beaver Dam 130
Beaver River 4, 15, 159
Beavercrow 160-161
Beavers 111-112
Birch River 130
Boiler Canyon 4, 148-150, 161
Boiler Rapids 72
Boulder Bar 155
Bradford, Leland (Wink) 9-10, 19-20, 29, 33, 56-57, 62, 66, 88, 95, 97, 101-102, 107, 117, 123, 128, 136-138, 145, 148

Brock, Thomas L. 2, 71, 76
Brown, Mr. 144
Bruce, J.B. 35
Brule, Etienne 48-49
Buffalo 54-55
Burnt Rapids 72

Campbell, Robert 35-36, 38-46, 69, 74, 78, 87-88, 161
Camsell, Charles (Dr.) 2, 12, 74-76, 79, 81, 90, 131-132, 135-136, 161-162
Camsell, Fred 2, 12, 74, 84, 90
Canadian Geographic Society 12
Carey, Dan 74, 78-79
Carey, Willy 74
Carlton House 45
Cascade Rapids 72
Cassiar 15, 83-84, 88
Cassiar Mountains 58
Cedar Lake 54
Champlain, Samuel de 48-50
Chicago 30, 82
Chilkat Indians 44
Chinese 3, 7
Churchill River 54
Clearwater River 54
Coal Creek 83
Coal River 3, 62-63, 66, 80, 82, 161
Coast Mountains 136
Columbia River 54
Colville River 39
Coureur de bois 48, 50; see also Voyageurs

INDEX

Cranberry Portage 21, 25-34, 47-48, 56
Cranberry Rapids 3, 25-28, 34-35, 161
Crow River 105
Crow Wing (Minnesota) 45

Dams 159-162
Dawson City 15, 75-76, 81
Dawson, George Mercier (Dr.) 2, 11, 31, 36, 38, 161
Dease Lake 10, 25, 38-39, 41-43, 45, 69, 73, 75, 83-84, 136
Dease, Mr. 39
Dease River 2-3, 11, 13, 43, 80, 83, 87, 122
Deer River 137
Devil's Canyon 79
Devil's Gorge 79, 91, 137-138, 159-161
Devil's Pool 139
Devil's Portage 4, 31, 40-43, 78, 86-98, 136
DeWolfe, Mr. 77
Diary of an Expedition from Edmonton (book) 71
Edmonton (Alberta) 2, 15, 30, 71-72, 76, 81-83
Edmonton Bulletin 71-72, 82
Edwards, Mr. 82

Finlayson River 81
Fornelli, Tim 68
Fort Chipewyan 38, 45, 54
Fort Frances 44
Fort Garry (Winnipeg) 45
Fort Halkett 2, 4, 35, 37, 39-40, 42-44, 46, 78, 80, 87
Fort Highfield 42
Fort Liard 4, 13, 39-40, 43, 73-74, 82, 111, 114-115
Fort McPherson 72
Fort Nelson 90, 95, 102, 104, 134, 158
Fort Nelson River 4, 83, 95, 103, 107, 110, 134, 159
Fort Pelly 45
Fort Resolution 45

Fort Selkirk 37, 44-45
Fort Simpson 2, 15, 30, 35-39, 41-46, 54, 72-76, 90, 95, 115, 126, 130-131
Fort William 8, 51, 54
Fort Yukon 36-37
France, French 48-51; see also Voyageurs
Frances Lake 11-12, 25, 39, 43-44, 46, 54, 75, 81, 83
Frances River 3, 80
Fraser, Doug 9, 19-20, 22, 56, 66, 70, 93, 99-101, 106, 108-109, 117-118, 133
Fraser, Harvey ix, 7-11, 11, 13, 19-20, 22-23, 29, 32-33, 57, 59, 65-67, 88, 99-101, 103, 106, 117, 134, 136-138, 139, 145, 147, 157
Fraser River 54
Fraser, Simon 36
French River 49, 51
Frog Portage 54
Fur traders 2, 4, 8, 35-36, 38, 45, 48-51, 54-56, 58, 72, 91, 127, 158

Geological & Natural History Survey of Canada (book) 1, 7, 31, 56, 86
Geological Survey of Canada 2, 36, 136
Georgian Bay 49, 51
Gibney, Alec 76-81
Glenora 83
Gold 2-3, 12, 15, 30, 67, 70-85, 91, 148, 158; see also Klondikers
Graham, Mr. 77, 81
Grand Canyon ("The Gates") ix, 1, 4, 7, 40, 73-74, 78, 89-90, 94-95, 135-158
Grand Portage 53-54
Grand Rapids 72
Gras Cap 131
Grave sites 15-16, 53
Grayling (fish) 9, 14
Great Lakes 49, 49, 51, 53
Great Slave Lake 45, 54
Grizzly bears 93, 98, 128, 139; see also Bears

Grundy, Edward 82
Hambleton, Dale 9, 14, 21, 29, 61, 66, 93, 98-99, 119
Hanbury, David 136
Harbottle, Jeff 9-10, 19, 32, 62, 66, 93, 95, 99, 117, 123, 128
Harris, W.H. 84
Hell's Gate (Hell Gate) 1, 4, 40-42, 78-79, 90, 105, 150-153, 155, 161
Hemming, Arthur 71
Highland River 3
Hoeltschi, Frank 10, 19-20, 28-29, 32, 58-59, 61-62, 66, 93, 105, 110-111
Hoole Canyon 81
Hoole, Frances 41
Hopwood, Victor G. 50
Hornets 66-67, 93
Hotsprings 67-68, 79, 88, 135; see also Liard Hotsprings Provincial Park
Howey, Bill 76-77
Hudson's Bay 47
Hudson's Bay Company (HBC) x, 2, 4, 25, 31, 36-37, 39, 41, 43, 45, 52, 72-73, 82-83, 87, 91, 111, 114-115, 122, 131-132, 144, 149, 162
Huron Indians 49
Hutchinson, Mr. 39, 41
Hutton, Mr. 82-83
Hydroelectric development, see Dams

Ile-a-la-Cross 45
Illinois River 50
Indian Summer 140, 143
Indians, natives 13, 30, 35, 37, 39-42, 44, 46, 48-49, 69, 73-74, 78, 80, 82, 103, 112-114, 122, 128, 132-133
Isle de Grave's 79
James, William 162
Johnson, Nels 84
Kechika River 3, 58, 82, 161
Kennedy, Mr. 81

Kitza (Indian) 41
Klondike creeks 71
Klondike Rush Through Edmonton (book) 15, 30, 73, 81
Klondikers 2, 11-12, 15, 30, 70, 72-73, 81, 86, 161-162

La Biche River 82
La Loche river 47
La Pierre House 37, 44
Lachine (Montreal) 49, 51, 53
Lake Nippissing 49, 51
Lake of the Woods 54
Lake Winnipeg 54
Lapie (Indian) 41-42
Lee, Alfred E. 2, 71, 76-81, 161
Lepine, W. 122
Lewes River 44
Liard Canyon, Upper Liard Canyon 3, 11-12
Liard Crossing 26, 88
Liard Hotsprings Provincial Park 4, 88
Liard Range (mountains) 123
"Liard River Hydroelectric Development" (report) 159
Liard River, see references throughout
Liard Spire 142-143, 148
Little Canyon 1, 3, 7, 14-22, 31, 161
Long Rapids 72
Lower Liard 4
Lower Post 3, 10, 12-13, 73, 80, 83
Lynn Canal 44

MacGregor, J.G. 15, 30, 73, 81
Mackenzie, Alexander 36
Mackenzie District 37, 39, 45, 122, 131
Mackenzie Mountains 123, 128
Mackenzie River 1-2, 5, 36, 39, 43-45, 54, 72-74, 76, 78, 123, 128-129, 131, 136, 153
Mattawa River 49, 51
McConkey, Jim 135, 147
McConnell, R.G. x, 1-2, 6, 11, 14, 17-18, 25-28, 31-32, 56, 58-60,

INDEX

66, 69, 86, 91, 93-94, 105, 122, 132, 136, 141, 144-146, 152, 161-162
McCulloch, Alex 84
McDame Creek 80
McDonald, Archibald 47, 52
McFarlane, D.G. 160
McLennan, Hugh 48, 50
McLeod, John M. 35, 37-39, 161
McLeod, John (Senior) 37-38
McLeod Portage 47
McPherson, Mr. 43
Methye Portage 54
Middle Rapids 72
Mining, Minister of Mines 2, 74; see also Gold; Klondike
Mississippi River 1
Missouri River 50
Montreal 45, 49, 51, 54
Moose 67, 87, 93-94, 98, 104-107, 109, 112-114, 128, 134, 139, 148, 161
Mountain goats 128
Mountain Portage Rapids 3, 56, 58-59, 161
Mountain Rapids 83
Mountain sheep 128
Mowat, Mr. 35
Mud River 56, 58
Mud River Post 3, 82
Muskoka region (Ontario) 8
Muskwa River 102

Nahanies (Indians) 42-43
Nahanni (book) 127
Nahanni Butte 5, 122-130
Nahanni National Park 128
Nahanni River 5, 123-128
Nelson Forks 15, 110
Nelson, Knute 82-83
New Caledonia 46
New York 48-49
Norman (fort) 46
North West Company 131
Northwest Territories 1
Norway House 37

Okanagan 74

Oliver, Mr. 82-83
Ontario 8, 48-49, 55, 150
Oregon Trail 50
Orkneymen 50
Ottawa River 49, 51
Otter 111

Pacific Ocean 1, 41, 43, 47
Parkman, Francis 50
Peace River 1, 54, 161
Peace River: a Canoe Voyage from Hudson's Bay to Pacific (book) 47, 52
Peel River 37, 44
Pelly, A.M. 74, 76-81
Pelly Banks 25, 44, 46, 73
Pelly, H. (Sir) 44
Pelly Mountains 1
Pelly River 41, 44, 73, 81
Petitot River 4, 73, 114-115
Police, RCMP 12-13, 114-115, 132
Porcupine River 44
Portage Brule 3, 80, 82, 85-86, 161
Portage Brule Rapids 39, 41, 63-64, 68-69
Portages, see Cranberry; Devil's; Frog; Grand; McLeod; Methye; Mountain; Portage Brule
Porter, James 83
Proposition Island 77
Purgess, G. 76

Quebec 48-49, 55; see also Montreal; Trois Rivieres; Voyageurs

Rabbit River 3, 59
Rabbits 102-103
Rapids of the Drowned 4, 72, 79, 105, 143, 145-146, 149, 161
Rapids, see Boiler; Brule; Burnt; Cascade; Cranberry; Grand; Long; Middle; Mountain Portage; Mountain; Portage Brule; Rapids of the Drowned; Whirlpool
Rat River 72
Red River 45

Report on an Exploration in the Yukon District... (book) 36
Rivers of Canada (book) 48
Riviere des Vents 86-88, 105, 161
Rocky Mountains 4, 41, 63, 91, 140, 153
Royal Canadian Mounted Police, see Police
Russia, Russians 39, 42-43, 69

Saskatchewan River 54
Schreeves, Mr. 81
Scotland, Scottish 39, 45
Shakes, Chief 42
Shand, Mr. 81
Shaw, Mr. 77
Shrum (G.M.) Generating Plant 161
Silvester's Post 80
Simpson, George (Sir; Governor) 38-39, 41, 43, 47, 52, 80; see also Fort Simpson
Skeena River 1
Slave River 54, 72
Smith River 2, 4, 37, 39, 80, 87, 161
Snyetown 82
Son of the North (book) 2, 12, 74, 131, 135-136
St. Lawrence River 49-50
Stein, Mike 136, 148
Stephen, T.A. 76-77
Stewart, A.D. 81
Stewart River 44
Stikine River 1, 42, 83
Sturgeon-Weir River 54
Sulphur Creek 143
Surrender Island 4, 143, 161
Swan District 45
Swedes 49
Switzerland 10

Telegraph Creek 12, 71, 83
Terminal Range (mountains) 63
Teufele, Adolf 8-10, 20-21, 33, 61, 65, 66-67, 70, 89, 95, 97, 101-102, 118-119, 128, 130, 136, 147, 155
Thompson, David 36, 50

Thunder Bay 51
Toad River 4, 74, 77-78, 83, 95, 102
Travels in Western North America (book) 50
Trepanier, Mr. 122
Trois Rivieres 50
Trout River 4, 79, 82, 90, 95-96, 101, 161
Turnagain River 161
Turner, Dick 127, 130

United States, see Americans

Vancouver 10, 136
Velge, Monte 80-81
Voyageurs x, 2, 8, 11, 17, 27, 30-31, 35-36, 39, 44, 47-55, 57-58, 69, 72, 86, 96, 122, 136, 144, 161-162

Washington (State) 9
Watson Lake 3-4, 26, 95, 97, 103, 111
Wenger, Ferdi (author) see references throughout
Whirlpool Canyon 3, 59, 62
Whirlpool rapids 161
White, Fred K. 84
Wildlife 41, 43; see also Bears; Beavers; Buffalo; Grizzly bears; Hornets; Moose; Mountain goats; Mountain sheep; Otter; Rabbits; Wolves
Windy River, see Riviere des Vents
Wings of the North (book) 127
Winnipeg 45
Winnipeg River 54
Wolves 104-107, 134, 161
Wood Buffalo National Park 54
Woodward, Harry 76-78, 81
Wrangell (Alaska) 11, 83
Wright, D.W. 74, 76-80

York Factory 37, 52
Yukon 1-2, 10-11, 15, 36, 45-46, 54, 67, 70-71, 73, 75, 77-78, 85, 113, 122, 161; see also Dawson City; Fort Yukon; Klondike; Yukon River
Yukon River 1, 73, 81

PRINTED AND BOUND
IN BOUCHERVILLE, QUÉBEC, CANADA
BY MARC VEILLEUX IMPRIMEUR INC.
IN SEPTEMBER, 1998